WAKING THE GIANT

BILL McGUIRE

WAKING THE GIANT

How a changing climate triggers earthquakes, tsunamis, and volcanoes

OXFORD
UNIVERSITY PRESS

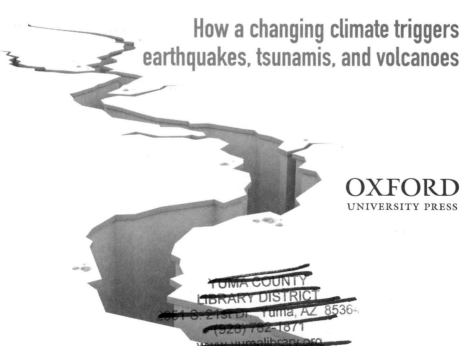

OXFORD

UNIVERSITY PRESS

Great Clarendon Street, Oxford OX2 6DP

Oxford University Press is a department of the University of Oxford.
It furthers the University's objective of excellence in research, scholarship,
and education by publishing worldwide in

Oxford New York

Auckland Cape Town Dar es Salaam Hong Kong Karachi
Kuala Lumpur Madrid Melbourne Mexico City Nairobi
New Delhi Shanghai Taipei Toronto

With offices in

Argentina Austria Brazil Chile Czech Republic France Greece
Guatemala Hungary Italy Japan Poland Portugal Singapore
South Korea Switzerland Thailand Turkey Ukraine Vietnam

Oxford is a registered trade mark of Oxford University Press
in the UK and in certain other countries

Published in the United States
by Oxford University Press Inc., New York

British Library Cataloguing in Publication Data

Data available

Library of Congress Cataloging in Publication Data

Data available

Typeset by SPI Publisher Services, Pondicherry, India
Printed in Great Britain on acid-free paper by
MPG Books Group, Bodmin and King's Lynn

ISBN 978–0–19–959226–5

1 3 5 7 9 10 8 6 4 2

CONTENTS

PREFACE

The 2004 Indian Ocean tsunami was the most shocking natural catastrophe of modern times, not because of the numbers of deaths and injuries involved, which were indeed mind-numbing, but because it reached into the lives of those convinced that they were safe, and destroyed them. By far the greatest number of casualties were residents of three developing nations, Indonesia, Thailand, and Sri Lanka, but because the waves crashed into popular tourist resorts, as well as obliterating native towns and villages, many thousands of casualties originated in industrialized countries from all parts of the world. Visitors from Preston in the UK, from Pittsburgh in the United States, from Perth in Australia, from small communities and great conurbations in countries as far flung as Sweden, Germany, Japan, and New Zealand, lost their lives, their limbs, their children, and their long-held feelings of immunity. Before 26 December 2004, natural disasters were things that happened to other people in lands far away; after that seminal day, we were all suddenly more aware of our fragility, both as individuals and as part of a global society, and of our vulnerability to Nature's heartless and random culling.

Regrettably, however, it does not appear to have encouraged us to try to improve our understanding of the planet we live on and how it functions. This is made depressingly clear by the continuing denial, in some quarters, of anthropogenic climate change. In the wake of the

2004 tsunami it also became apparent very quickly through the bizarre explanations for the catastrophe that proliferated at viral speed across the internet. For the especially ignorant, the terminally paranoid, or those with irrationally suspicious minds, nothing is ever as it seems. To these, the tsunami cannot have been simply a natural consequence of the realignment of the stresses and strains within the interior of our world, but must have reflected something deeper and more deceitful. Maybe the Chinese caused it, or was it the result of the US military testing yet another mind-bogglingly surreal secret weapon? Or, perhaps it was climate change? This last from a bunch of individuals in direct opposition to the climate sceptics camp, largely lacking a formal scientific education or, in some cases it seems, any education at all, who are able to convince themselves that they see the hand of anthropogenic climate change in almost every notable event, whether natural or unnatural. Even before the bodies were buried, the internet was crackling with theories speculating about how climate change might have played a role in the great catastrophe. Hardly surprisingly, most suppositions teetered on the edge of madness, while others were just downright wacky. Acknowledging the title of this book, none actually invoked a literal subterranean giant as the ultimate cause, but they might just as well have done, given some of the incredible explanations. My favourite of the lot confidently espoused the idea that, as a consequence of global warming, 'magma in the Earth's core [sic] is heating up, raising the Earth's temperature and causing eruptions and earthquakes'. This is wrong on so many levels that it is difficult to know where to begin; suffice to say that while warming is penetrating to deeper and deeper levels in the oceans, there is no possibility of it infiltrating the extra 3000 or so kilometres to our planet's core. Even if it did, what possible difference could a rise of a few degrees Celsius make to the Earth's metal heart, quietly roasting away at a temperature close to that of the surface of the Sun.

What probably attracted me to this, of all the weird and wonderful explanations for the Indian Ocean tsunami, was that it did contain a grain, if not of truth, then of acumen. The idea that climate change caused the Sumatran earthquake and resulting devastation is clearly nonsense, although this has not prevented the 'theory' rearing its ugly head once again in the wake of the 2011 Japanese tsunami. This said, however, it is becoming increasingly apparent that our world's climate has in the past interacted with the solid Earth (or geosphere) and the processes that occur therein, and continues to do so today. Both natural climate change and that arising as a consequence of human activities, tend to be thought of as involving solely the atmosphere and the hydrosphere (the oceans, rivers, lakes, glaciers, ice sheets, and the like), but the true picture is far more complicated. It is now more than 40 years since James Lovelock first began trailing the idea—known as the *Gaia* hypothesis—that our Earth was a complex, interactive system incorporating the atmosphere, hydrosphere, geosphere, and biosphere. It has been a long and hard struggle for Lovelock's core message—that life unconsciously plays a key role in maintaining climatic and biogeochemical conditions on our planet in a manner that keeps it suitable for life itself—to start to take hold and gain scientific credence; Gaia theory and its ramifications continues to generate vigorous debate, alongside unwarranted criticism and downright hostility from a few individuals who have not quite grasped what Lovelock is getting at. Even critics would agree, however, that the concept of Gaia has been pivotal in changing the way we view our planet, no longer as an assemblage of disconnected elements but as an integrated and interactive whole. In this regard, Lovelock's idea was instrumental in establishing the concept of Earth System Science, a way of observing and studying our planet that addresses the entire jigsaw rather than just its constituent pieces. Earth System Science has made it easier to think about and develop ideas that involve connections and interrelationships that might have been more difficult to speculate upon, without derision, in pre-Gaian times. These include interactions between climate and weather, and processes operating within, and at the surface of, the geosphere, including volcanic eruptions and earthquakes, leading to consequences that could prove disastrous in certain circumstances.

Without the change of approach that Gaia theory and Earth System Science have fostered, it is unlikely that more than 70 scientists would have felt happy gathering in 2009 at University College London to attend a conference entitled: *Climate forcing of geological and geomorphological hazards*, and even more unlikely that the Royal Society, the world's pre-eminent scientific organisation, should have felt relaxed about publishing papers presented at the meeting in its *Philosophical Transactions*. Today, however, the enormous amount of research that has gone into better understanding anthropogenic climate change—without any doubt the greatest threat our society has ever faced—has revealed all sorts of complex feedback loops, unexpected correlations and surprising interrelationships between different elements of the Earth System. This has helped the scientific community become far more open to new ideas and models provided, that is, that they are grounded in sound observation and well-argued interpretation. As a consequence, while the possibility of a climate change origin for the Indian Ocean tsunami remains confined firmly in the minds of the muddle-headed, the idea that future climate change may drive an increase in earthquake activity is accepted as one possible outcome of the changing climate educing a response from the Earth's interior.

At the risk of leaving myself wide open to accusations of anthropomorphism, maybe it is helpful to think of the broadly benign Earth beneath our feet in terms of a slumbering giant, which, in December 2004 stirred off the coast of Indonesia as a reaction to tectonic forces that could no longer be reigned in. Since then, the giant's tossings and

turnings have prompted another devastating tsunami in Japan, catastrophic earthquakes in Pakistan, China, and Haiti, lethal landslides in
the Philippines, Bangladesh, and El Salvador, and disruptive volcanic
rumblings in Iceland. A wish to unpick the ways and means by which
the giant has responded to serious environmental upheavals in the
past, particularly at times of major climate transition, and to shed light
on its likely riposte in a future world that may, by 2100, have changed
out of all recognition, provides the justification for this book.

This is not intended as a speculative rant based upon unattributed
hearsay, unfounded data, and unverifiable observation, but a straightforward presentation of what we know about how climate and the
geosphere interact, combined with informed discussion about what
implications such knowledge may hold for the future. Inevitably, some
conjecture is involved, but firmly grounded—I hope—in sound, peer-
reviewed science. As will become apparent from the chapters that follow, rapid, and dramatic environmental changes in the past, particularly
at times of major climate transition, have together elicited a considerable response at the Earth's surface and beneath. Many geological systems such as extant volcanoes, active faults, and unstable slopes are
shown to be often critically poised so that even tiny external perturbations can be capable of triggering reactions in the form, respectively, of volcanic eruptions, earthquakes, and landslides. In this light,
it would be surprising indeed if the melting ice sheets, rising sea levels,
and changing weather patterns that will undoubtedly characterize the
coming century and beyond, did not go some of the way—at the very
least—towards reawakening the slumbering giant beneath our feet.

Bill McGuire
Brassington, Derbyshire

August 2011

LIST OF ILLUSTRATIONS

The Storm
after the Calm

Many imaginative analogies have been devised in order to get across more accessibly the idea that the human race has 'graced' the surface of our planet for virtually no time at all—at least in comparison to how long the Earth has been around. According to one, if our world's entire history could be compressed into a single year, humans would appear on the scene just before 8 p.m. on New Year's Eve; civilisation emerging just 42 seconds before Big Ben's chimes. Another—one of mine actually—likens the history of our planet to a team of international athletes running a mile. Using this analogy, the explosion of complex life that marked the start of the Cambrian Period would not take place until well after the bell has rung and when the athletes are on the back straight of the final lap. Dinosaurs make an appearance as the runners battle for the lead in the home straight and disappear again with the leaders still 25 metres from the finish, while the human race finally pops up in the last split second of the race when the winner is inches from breasting the tape. Whichever way you choose to look at it, in comparison to the age of the planet we live on, it is quite apparent that we are the newest of new tenants on the block. Looking back over the unimaginably long history that preceded the arrival of modern humans, and into what is now fashionably known as 'deep time', it is impossible not be to be

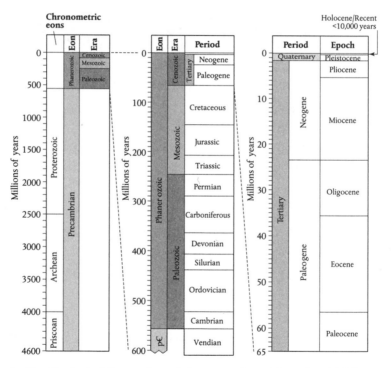

Fig 1. The Geological timescale usefully compartmentalises the 4.6 billion year history of our planet, but cannot convey the enormous span of time that has passed since our world built itself from bits and pieces of debris during the solar system's earliest days.

impressed by the batterings our planet has taken and the wild swings in climate it has experienced—and by the fact that is has always *it* bounced back. On a number of occasions it looks to have been out for the count, only to haul itself to its feet and rebuild an even more astonishing assemblage of organisms.

Down but not out

It is not my intention to provide here a comprehensive history of our planet, or even a potted one. In order to appreciate just how hostile an environment the Earth can drum up, however, it is worth taking a look at a few of the more unsavoury episodes that have punctuated our world's existence to date, if only so that we can value the relative tranquillity of modern times. Like all the other bodies in our solar system, the Earth appeared on the scene a little more than 4.6 billion years ago—a staggering length of time, but still more than nine billion years after the Big Bang that brought our universe into existence. Rather like an out-of-control snowball rolling down hill, our planet built itself by means of sweeping up debris that filled local space at that time, a mechanism known as planetary accretion. A thin crust formed rapidly, despite the planet continuing to take a pounding from the many chunks of rock that still wandered the solar system, and may well have managed to spawn bodies of surface water and a primitive atmosphere. According to the currently favoured view, however, a very big shock was in store, arguably the greatest in our world's long history. Even before it reached its 100 millionth birthday, it seems likely that the Earth was struck a glancing blow by another celestial body, just a little smaller than Mars. This object, sometimes going by the name of Theia—the Titan that in Greek mythology gave birth to Selene, the Moon goddess—is charged with gouging out a sizeable chunk of our planet's crust and mantle, forming a ring of debris that eventually pulled itself together in orbit around our planet, along with some remains of Theia, to form the Moon. This cataclysmic event would have shattered the relative calm of the early Earth, obliterating the just-formed crust and replacing it—at least according to some—with a planet-wide magma ocean.

The collision also had ramifications that persisted down the eons and that are still important: most particularly the creation of the lunar tides that tug at our world's oceans and crust. There is even speculation that the collision was responsible for knocking the Earth off its axis of rotation, so that this was no longer vertical with respect to the plane of its orbit about the Sun, but tilted at an angle that ranges cyclically from a little over 22° to about 24.5°, over the course of 41,000 years or so. This may not sound a big deal, but it has fundamentally affected our planet's climate, not only resulting in the seasons, but also playing a critical role in the climatic cycles that control the timing of the ice ages. As we shall see in later chapters, the ice ages and the seasons, together perhaps with the lunar tides, have important roles to play in triggering or regulating geological events even today. If our world's skewed axis of rotation really is a consequence of the body-blow that the Earth experienced in its formative years, then the collision's shuddering consequences have prevailed down more than four billion years, and continue to make themselves felt.

While this earliest cataclysm had the most profound ramifications of any in Earth history, a number of not insignificant shocks were to follow. After a couple of billion years of cooling—during which the crust stabilized, plate tectonics became established and simple life forms appeared—our world suffered the first, and most severe, of its great ice ages. The Huronian glaciation covered large areas of the planet between around 2.4 and 2.1 billion years ago; its initiation seemingly triggered by the activities of early life. By this time, tiny

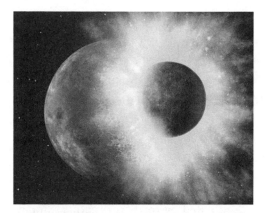

Fig 2. The collision between the Earth and a Mars-sized rogue object came close to splitting our world apart. The remnants of this cataclysmic crunch came together to form the Moon and may have tilted the Earth off its axis, leading to consequences that are still apparent today.

organisms known as cyanobacteria had been successfully photosynthesising—combining carbon dioxide and water from the Earth's atmosphere and surface, catalysed by sunlight, to produce sugars and oxygen—perhaps since as far back as 3.5 billion years. For maybe a billion years or more, these lowliest of life forms thrived on the sugars produced in the process, while the oxygen released as a by-product was captured by the oceans, through chemical reactions with sediments and organic matter and by the formation of the planet's protective ozone layer. Just prior to the Huronian glaciation, however, all these so-called oxygen 'sinks' were filled to overflowing, such that 'free' oxygen started—for the first time in Earth history—to accumulate in the atmosphere. This momentous event, without which neither we nor most of the rest of the great tapestry of terrestrial life would exist, is known as the Great Oxidation Event or GOE, and sometimes as the Oxygen Crisis or the Oxygen Catastrophe.

Why catastrophe? That's because this was just what it was for the countless organisms then living, for which the appearance of free oxygen was far from something to celebrate. Most of Earth's inhabitants at the time were micro-organisms that respired anaerobically, that is, without the involvement of oxygen, and for the majority of these the growing quantities of oxygen in the atmosphere of our planet meant poisoning and death. So these early oxygen-producers brought about a transformation in the biosphere of the time, annihilating most of the life that eschewed oxygen and setting the scene for a future world dominated by oxygen breathers, including ourselves. A second outcome of the GOE was to bring about a drastic change in the planet's climate, as oxygen combined with methane—then a common constituent of the atmosphere—to form carbon dioxide and water. Mysteries remain, but it appears that this wholesale removal of methane—a potent greenhouse gas—reduced the effectiveness of the blanket of atmospheric compounds that had kept the Earth relatively warm over the previous few billion years, resulting in plunging temperatures and the widespread growth of ice sheets across much of the planet.

More than a billion years after this earliest age of ice, our world was once again plunged into the deep freeze, as dramatic cooling returned. While the cause is not certain, it is likely that, for one reason or another, this second frigid episode involved a reduction in the concentration of another greenhouse gas in the atmosphere, this time carbon dioxide. The result was the Cryogenian, a period that lasted from around 850 to 635 million years ago, during which virtually the entire Earth seems to have been enclosed—at least once—in an icy carapace: great ice sheets meeting at the equator, where temperatures were as low as in Antarctica today. While the bitter conditions of so-called Snowball Earth did life no favours, it took little time for it to recover and thrive once greenhouse gases from the world's active volcanoes eventually started to warm the planet once again. Indeed, evolutionary pressures during this time of dramatic global climate change may have provided the trigger for the explosion of multi-cellular life that characterized the Cambrian Period that followed a hundred million or so years later.

Over the past half a billion years or thereabouts, life on Earth has become increasingly complex, culminating, during the past 15 million years, in the appearance of the great apes (which today include the gorillas, chimpanzees, and orang-utans as well as ourselves) and, around 200,000 years ago, modern humans. During this period, however, a number of natural calamities occurred that ensured that the growth of the tree of life was anything but straightforward. Five mass extinctions, during which more than half of all animal species were wiped from the face of the Earth, provide eloquent testimony to this. Following on the heels of two less formidable extinctions in the previous few hundred million years, the greatest culling was undoubtedly the 'Great Dying' that occurred at the end of the Permian Period about 250 million years ago, and which involved the loss of virtually all marine animal species and nearly three-quarters of land animal species. Theories for this 'mother of all extinctions' abound, and include a general deterioration of environmental conditions; sudden climate change driven by a vast release of methane from sea-floor sediments; a massive comet or asteroid impact; and the largest outpouring of lava in Earth history. While the jury remains out, it is hard—especially as a volcanologist—not to be swayed towards the volcanic hypothesis by the unimaginable scale of the rumblings going on in Siberia at the time. Immediately prior to the mass extinction, this region hosted the disgorging of at least three million cubic kilometres of lava, burying an area equal to more than half that of the USA to depths of more than three kilometres. These unprecedented effusions were accompanied by colossal quantities of ash and volcanic gases with serious climate modifying potential. Whatever the reason for the wholesale slaughter at the end of the Permian, it was not the last of the catastrophes to punctuate the relative calm of our evolving world. Another somewhat less severe extinction occurred 50 million years or so later, at the end of the Triassic Period, attributed to one or other of the usual suspects, and clearing the decks for the age of dinosaurs. Approaching 150 million years later, the reign of the giant reptiles was suddenly, and almost certainly violently, curtailed in the best known and most hotly disputed mass extinction of all. A little more than 65 million years ago, at the end of the Cretaceous Period, the so-called K-T Extinction Event (or, as it is now officially known, the K–Pg Extinction Event) heralded the demise, not only of the dinosaurs, but also of more than three-quarters of all animal species. Other possibilities have been put forward, but two main candidates head the running as potential triggers of this biological mayhem: a 10 km asteroid or comet impact off the coast of Mexico, which has left as its legacy the 180km-wide Chicxulub crater; and the voluminous Deccan Trap flood basalt outpourings in western India, which occurred at around the same time. Debate and dispute continues as to which was responsible, but a 2010 appraisal in the journal *Science*, by more than 40 scientists, comes down firmly on the side of the impact scenario, putting this theory in pole position—for the moment at least.

While the K–T (or K–Pg) event marked the youngest of the great extinctions, other knock-backs continued to happen, albeit on a smaller scale. Two of 15 lesser extinction events identified in Earth's long history are recognized after the end of the Cretaceous, one terminating the Eocene Epoch about 40 million years ago and another taking its toll around 25 million later during the Middle Miocene. Along with these incidents which, like the major extinctions, mark abrupt changes to the environment due to either terrestrial or external influences, the Earth's climate also started to show signs of a more profound change. Slowly but surely our planet was starting to cool as it travelled a road that would result ultimately in the relentless march of the glaciers across much of Europe and North America. About 10 million years ago, after an earlier abortive attempt, ice sheets started to get a grip in what had been an ice-free Antarctica. Cooling soon affected the northern hemisphere too, and by five million years ago, both the great southern continent and Greenland were buried under sheets of ice, probably as thick as those we see today. It seems that an unsuccessful attempt was made by the Arctic ice to expand southwards at around about this time, but the Ice Age proper did not start until around two and a half million years ago. Since then, the ice sheets have ebbed and flowed across the land masses of the northern hemisphere with eight cycles of cold and warmth recognized over the past three-quarters of a million years alone. The hyperactive Quaternary Period, during which all this happened, is examined in more detail in the next chapter. Suffice it to say here that our world remains embedded in a sequence of freeze and thaw that, despite the warming effects of human activities, will see the ice return again some time in the future.

The rise and rise of the human race

It is now just 20,000 years or thereabouts since the ice sheets reached their greatest extent, at the time of the Last Glacial Maximum or LGM. In the intervening millennia our world has been largely transformed from one reminiscent of the endless winter of C. S. Lewis's *The Lion, the Witch and the Wardrobe* to one of green and pleasant lands, tropical idylls, and wide oceanic vistas characteristic of our warm and watery interglacial. As we shall see in later chapters, the switch was not a smooth one and for much of the immediate post-glacial period our planet was held deep in the throes of an unruly Jekyll and Hyde transformation from ice world to water world. It is not easy to imagine that barely 800 human generations ago, much of the northern hemisphere was a frozen wasteland. Kilometres-thick ice sheets extended far to the south across Europe, North America, and parts of Asia, their formation sucking up the planet's water so that global sea levels were as much as 130m lower than today—a whisker less than the height of the London Eye Ferris wheel. Ice-world temperatures averaged a good five degrees Celsius lower than in the balmy world we have come to know and love. The following 20 millennia, however, saw a remarkable change, as climbing temperatures forced the rapid decay of the continental ice sheets, pouring meltwater into the ocean basins and driving up sea levels that at times were mounting at a rate of several metres a century. Land bridges were swamped, new seas were formed and atmospheric circulation patterns changed to accommodate broadly warmer, wetter conditions, leading to modification of major wind trends and a rearrangement of climatic zones.

By around 12,000 years ago things had settled down pretty much, heralding a period of relative climatic and geological calm known as the Holocene, and setting the scene for the meteoric rise of human civilisation. In the face of aforementioned evidence for the Earth's terrifyingly dynamic nature, surely it can't be coincidental that our race has transformed itself over a dozen millennia from a million or so Stone Age hunter-gatherers to a technology-driven global society of seven billion people during a long and sustained period of environmental stability. Without the broadly benign climate of the past few thousand years, together with the fortuitous absence of volcanic super-eruptions and collisions with large asteroids, is it at all likely that we would have been be in a position, in 2009, to switch on the Large Hadron Collider, in our efforts to unravel the universe's deepest and darkest secrets? Almost certainly not. Appreciating this fact is critical, because it is becoming increasingly apparent that this benevolent, but transitory, episode of recent Earth history is rapidly drawing to a close. Its termination will arise not as a consequence of collision with an object from space, nor from a volcanic cataclysm, whose frequencies are measured in tens or hundreds of thousands of years, but as the upshot of our own activities.

Human civilisation developed and thrived during a fortuitous lull following the end of a post-glacial storm of climatic and tectonic dynamism. It looks almost certain, however, that this golden age will very shortly succumb to the fury and carnage of yet another tempest—this time, one of our own making. Prior to the Holocene, our world's stormy history was a reflection of natural change, often driven by processes operating within and upon the Earth, sometimes involving life itself, and occasionally promoted by extraterrestrial intruders such as asteroids and comets and, in its earliest days, by our Moon's progenitor. Even as the ice sheets were in retreat, however, a new force for change appeared, to accompany nature and, in time, to overwhelm it. As the world warmed and the ice faded, the climate proved increasingly antagonistic to a wide range of large mammals. Hardly surprisingly, many that were attuned to a world of ice and snow found the new conditions especially difficult. Numbers of mammoths and mastodons went into steep decline between 15,000 and 10,000 years ago and most members of both groups soon after became extinct, joining the ranks of the sabre-toothed cats, the ground sloths, the cave bear, and the Irish elk. Despite the dramatic changes in their environment, however, just why the mammoths should have vanished at this time has always been considered somewhat problematical. After all, these spectacular giant elephants had been around for almost five million years and had managed to hang on throughout several earlier interglacial episodes. Certainly a warmer climate would not have been conducive to their survival, and habitat changes that saw grasslands and open woodlands replaced by forest would not have helped either, but these changes had happened before, and on their own they would probably not have been sufficient to bring about the disappearance of these majestic beasts from the face of the Earth. What made this transition from cold to warm different, however, was the existence of well-organized bands of human hunter-gatherers who sought out and brought down mammoths and other larger mammals for food, tusks, and skins. Evidence for spear use in hunting goes back more than 15,000 years to a time when mammoths and other elements of the large mammal megafauna were demonstrably in decline. It is not unreasonable to suppose, then, that the end of the mammoth's five million year reign across the ice fields and the tundra was brought to a conclusion through the conspiracy of a changed environment and pernicious hunting by humans.

In 2010, this idea was taken one step further when Christopher Doughty and colleagues at the Carnegie Institution for Science in Stanford, California, proposed that the extinction of the mammoths led to measurable climate change. Doughty and his fellow researchers discovered, through pollen analysis, that as mammoth numbers started to fall dramatically, so the small trees kept in check by munching mammoth teeth, most particularly birch, started to flourish and spread rapidly. As birch woods replaced grasslands across the wastes of North America and Siberia, so the reflectivity—or albedo—of the surface also changed, the darker tree cover absorbing more solar radiation than the lighter-coloured grasslands. According to the Carnegie scientists, this would have driven up temperatures across Siberia by 0.2°C, with rises as high as 1°C in places. In their own right, such temperature changes are small, and would have been only a minor contribution to the general warming of the planet during post-glacial times. Crucially, however, if humans had a hand in the demise of the mammoths, this provides the earliest evidence for the activities of our race influencing the climate. It marks out the early post-glacial period as the time when humans first interfered, in a wholesale manner, with their environment, and the time when human activities first contributed to global warming.

And this was just the start. Wind the clock forward several thousand years, and the hunter-gatherers that saw off the mammoths and other large mammals gradually voted with their feet for a lifestyle change that saw the rigours, dangers and uncertainty of foraging and hunting replaced by a quieter, less active and more secure existence. During the calm of the early Holocene, around 8,000 years ago, our ancestors started to abandon their hunter-gatherer lifestyle, settling down in communities to farm crops and husband animals, initially in the Middle East's 'fertile crescent', but very quickly right across the planet. Not only did this switch from wanderer to stay-at-home provide the template for our current, largely urban existence, it also had implications for the global environment and for the composition of the atmosphere that were far more fundamental than those attributed to the earlier mammoth hunters.

In a seminal paper published in 2003, Bill Ruddiman of the University of Virginia, advocated the idea that the birth of modern agriculture around eight millennia ago coincided with the first significant warming of the planet as a consequence of human activities. Ruddiman drew attention to the unusual trends in atmospheric carbon dioxide and methane that distinguish the current interglacial period from those that went before. Based upon records of gas concentrations from past warm periods, carbon dioxide levels would have been expected to peak around 10,000 years ago and then fall steadily. This they did, but only for a few thousand years. Around 8,000 years ago, carbon dioxide levels started to climb again—something not seen in any of the previous three interglacials—eventually reaching a value of around 280 parts per million, which was sustained until the industrial revolution. The methane picture is comparable. Methane concentrations follow a natural cycle that would normally be expected to have peaked around 10–11,000 years ago, as indeed they do, and then to fall. This again they do, but only until 5000 years ago. Then, following the pattern of carbon dioxide levels, they start to creep upwards once more, reaching levels, by the late Holocene, far higher than can be naturally explained.

Ruddiman's thesis is that both anomalous trends are a reflection of the hand of Man and the increasingly fundamental modification of the global environment as a result of human activities. The climbing carbon dioxide levels, Ruddiman holds, are a consequence of large-scale, unsustainable, deforestation across Europe and Asia, which started around 8000 years ago. Wood burning returned carbon dioxide locked up in the trees to the atmosphere while a massive reduction in forest area meant that less carbon dioxide was absorbed. In a similar manner, the timing of the rising methane trend coincides with the start of efficient, irrigated, rice production in Asia. Rice fields today constitute one of the biggest sources of atmospheric methane, which is an especially potent greenhouse gas. With the world's population during prehistoric times a tiny fraction of what it is today, perhaps just five million at the time of the explosion in rice production, it might be expected that even our ancestor's best efforts at environmental degradation and land-use change would not be sufficient to have a detectable impact on the climate—but this looks to be far from the truth. Ruddiman estimates that these early human activities raised global average temperatures by around 0.8°C, and up to 2.0°C at high latitudes. These figures are comparable with the effect that two centuries of industrialisation has so far had on the world's temperature although, of course, there is almost certainly far more to come. If anything, the ability of tiny populations of humans to measurably warm the planet provides a salutary warning in terms of the sensitivity of our world's climate to the activities of our race. If a few million primitive individuals can have a significant impact on the Earth's temperatures, imagine what an industrialized society of seven billion people can eventually accomplish.

Richard Arkwright's legacy

In 1732, a baby was born in the town of Preston in north-west England, a place steeped in the textile industry since medieval times. As the youngest of 13 children in a terribly impoverished family, Richard Arkwright had prospects close to zero. Overcoming his humble birth, however, Arkwright not only became one of the great entrepreneurs of his time, but also transformed the world. Arkwright's legacy is guarded by the Arkwright Society at Cromford Mill, just a few miles from where I live. The mill and others like it on the banks of Derbyshire's Derwent River, form part of a UNESCO World Heritage Site, because it is here, as much as anywhere, that the Industrial Revolution can be said to have started. It is here, in other words, that anthropogenic climate change really started to take off, leading, almost 250 years later, to our current predicament: a rampant, consumer-driven industry, forcing up greenhouse gas concentrations in the atmosphere and heating the planet at an ever-increasing rate.

As the leading entrepreneur and businessman that he was to become, Richard Arkwright would undoubtedly have been keen for his name to go down in history, but it is extremely unlikely that he could have had even an inkling of the impact he would ultimately have on our society, our planet, and everything living thereon. Following an early career as a barber and wig-maker, Arkwright used funds he had accumulated, as a consequence of a waterproof dye he had invented for wigs, to develop his interest in the cotton industry. Observing that producing yarn by hand using a spinning machine known as a Spinning Jenny could not keep up with growing demand, in 1769, at the age of 37, Arkwright invented the Water Frame, a water-powered device designed to produce stronger thread for yarns, more quickly and more efficiently than could be accomplished by its hand-powered forerunner. These machines were large and, as the name gives away, were powered by water. Consequently, they could not be operated in the homes of weaver families, as could the far more conveniently sized Spinning Jenny, and were instead installed in large buildings and powered by water wheels. These new installations looked just like water mills and came to be known simply as mills; the particular buildings where the water frames were housed were known as factories. Weaving families in the village of Cromford where Arkwright, along with impressively named partners Jedediah Strutt and Samuel Need, opened the world's first water-powered mill in 1771, saw their lives change overnight. Formerly, the women spun the yarn while their men-folk weaved the cloth, all within the confines of small cottages. Now, the men remained behind to weave, while the women had to head for the factories to spin the yarn. Not only had mass production arrived on the scene, but so had commuting. Arkwright even developed an efficient system for getting the most out of his workers: bells rang at 5 a.m. and 5 p.m. marking the beginnings of 13-hour overlapping shifts on which—at least to start with—children as young as seven toiled.

Cromford Mill is emblematic of a change in the way humans lived their lives that is comparable in scale and significance to the switch from hunter-gathering to a sedentary, agrarian economy, 8000 years earlier. Within decades, the concept of the mechanisation of work, which had previously been undertaken by individuals or small groups using just the power of human muscle, spread like wildfire. Arkwright himself rapidly expanded his empire, building further mills on the River Derwent and others in his home county of Lancashire and as far away as Bath and Scotland. The advent of steam power soon made the need for factories to be built close to watercourses redundant and ensured that mechanisation could cross over to other industries such as glass-making, chemicals, and machine tools, the last of which accelerated the revolution in work and production. Within decades, a British economy, previously built entirely upon manual labour, was galvanized and transformed. Coal refining, iron-making, textile production and a host of other processes embraced machine technology powered by steam. A profusion of new roads, canals, and railways carried the revolution into every corner of the country. In the century that followed, industrialisation proliferated rapidly throughout western Europe and North America in a seemingly unstoppable march across the face of the planet that continues today.

Arkwright's legacy then, is nothing less than the industrialisation of our world. Thanks indirectly to this 18th century businessman and inventor, 2009 saw human activities pump more than 30 billion tonnes of carbon dioxide into the atmosphere, the by-product of a technology-based global society that produced close to 70 million cars and commercial vehicles and more than 100 million bicycles; a society that flew not far short of five billion passengers from place to place; and that provided the infrastructure for two billion people to use mobile phones and a quarter of the world's population to access the internet. How could Arkwright have known, nearly 250 years ago, that his invention would kick-start a technological transformation that would see information and communication providers adding 600 million tonnes of carbon dioxide to the atmosphere so that we can transmit more than 2 million tweets a day and nearly a hundred trillion emails every year.

The composition of the Earth's atmosphere cannot have held any meaning for Arkwright, nor for most of his contemporaries, but for us it is at the core of one of the great issues of our—or any other—

time. When Arkwright and his partners proudly opened the world's first factory 240 years ago, the concentration of carbon dioxide in the clear, cool air of the Derwent valley was around 280 parts per million (ppm). This is more or less what might be expected in the middle of an interglacial period, and pretty much the same as in previous warm periods sandwiched between glacial episodes. Two and a half centuries of booming industrialisation, fed by the unrestricted burning of fossil fuels, uninhibited cement production and the ceaseless destruction of forests, have pushed this up by 40 per cent to 391 ppm. To satisfy the increasingly insatiable demands of unbridled consumerism, we have dumped into the atmosphere in 10 generations or so, the accumulated carbon of tens of millions of years, without, until very recently, any thought for the impact or consideration of the consequences—this despite the fact that the greenhouse gas properties of carbon dioxide have long been recognized and the outcome of increasing the concentration of the gas in the atmosphere long ago predicted.

The story so far

The idea of human activities heating the planet was first flagged up by Swedish physical chemist and Nobel Laureate, Svante Arrhenius, in the early years of the 20th century. Having seen the global warming problem coming for more than 100 years, then, it seems quite incredible that we have yet to act decisively in order to do something about it. Or maybe not so extraordinary. Humans, as individuals, as groups, and together as a society, seem to be hard-wired to respond quickly and effectively to a sudden threat, but not to a menace that makes itself known stealthily and over an extended period of time. During prehistory such a response was clearly beneficial and useful, perhaps in avoiding the trampling feet of a herd of stampeding mastodons, but it makes us poorly placed to cope with 'long emergencies' such as anthropogenic climate change, wholesale ecosystem degradation or the gradual depletion of hydrocarbons and other natural resources. Following the Japanese attack in 1941 on the Pearl Harbour naval base in Hawaii, the US economy completely transformed itself in just six months, retooling to fight a war against fascism in Europe and imperialism in the Pacific. Despite our increasingly desperate predicament, climate change has not prompted anything like this sort of response, and initiatives designed to cut carbon emissions, such as the Kyoto Protocol, have made no impression at all on the steadily rising concentrations of greenhouse gases in the atmosphere.

With no short, sharp, shock to galvanize us into action, our society's response to a progressively changing climate is akin to a runner hurtling flat out towards the edge of a precipice. Rather than slow down before reaching the rim, our choice—if we think about it at all—is to sail over the edge at speed, clutching at a finger of rock just before we plunge into the abyss, and hoping that it holds until we can pull ourselves back up again. Unfortunately, there is no guarantee that we ever will, and even if we manage to, it will be a long and gruelling climb.

Anthropogenic greenhouse gas emissions and the world's climate have evolved in tandem over the past century or so. As is now clear, the influence of changes in human behaviour on the climate during prehistoric times is detectable in the form of hikes in greenhouse gas concentrations in the atmosphere or in temperature. Not unexpectedly, with industrial expansion going full tilt, greenhouse gas levels have been climbing at ever faster rates since the start of the 20th century, reflected in climbing temperatures that are both predicted and difficult not to notice. According to a paper published in *Science* in 2009 by Aradhna Tripati of the University of California, Los Angeles, and colleagues, atmospheric carbon dioxide levels are already higher than at any time in the past 15 million years or so. Over the intervening period, concentrations were not drastically different from those encountered prior to the onset of industrialisation a couple of centuries ago, and we have to go back to the middle part of the Miocene Period—a time known as the Mid-Miocene Climatic Optimum, or MMCO—to find levels in excess of today's 391 ppm. This really puts the current situation into perspective. In the course of just a couple of hundred years, human activities have done more to change the greenhouse gas composition of the atmosphere than natural processes extending far back into the geological record have managed. One sobering thought, which I will return to in the following chapter, is that the carbon-rich atmosphere of the MMCO was accompanied by hot conditions, with global average temperatures 3–6°C higher than at present, and by sea levels hovering between 25 and 40 m higher. Tripati and her co-workers are not the only climate scientists who look back at the world of the Mid-Miocene and see there the future of our planet.

The inexorable rise in global average temperatures that has accompanied rising greenhouse gas concentrations in the atmosphere has become increasingly obvious in the last couple of decades, and is captured in the now iconic—though climate change deniers would say notorious—hockey-stick graph. Arguably the best known graph in history and perhaps the only one recognisable to at least some non-scientists, the term 'hockey stick' was coined by US meteorologist and climate scientist, Jerry Mahlmann, although the graph itself was compiled more than 10 years ago by climatologist Michael Mann, currently the director of Pennsylvania State University's Earth System Science Center, and fellow researchers. It shows global temperatures over the past thousand years as relatively stable (the shaft of the hockey stick), before climbing rapidly around the turn of the 20th century (the blade of the stick). The graph is sewn together from modern temperature records going back to about 1850, contemporary historical records and so-called proxy data from ice cores, tree rings and corals, that provide indirect measures of temperature. As such, it is a composite construct that incorporates data for which both accuracy and precision may vary, which is clearly less than ideal and which leaves the graph open to criticism and disparagement. Notwithstanding attempts by the climate change deniers to belittle its accuracy and worth, revelations about arguments within the climate science community at the time the graph was first drawn up, and some honest technical objections by fellow climate scientists, the message that the 'hockey stick' reveals is essentially sound: that our world

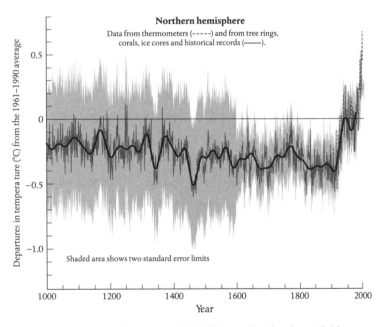

Fig 3. Perhaps the best known graph in history, the 'hockey stick' was compiled by climatologist Michael Mann and fellow researchers. It shows global temperatures over the past thousand years as relatively stable (the shaft of the hockey stick), before climbing rapidly around the turn of the 20th century (the blade of the stick).

is warmer now than at any time in the past millennium, and that the hike in planetary temperature over the past hundred years is unprecedented over that period.

According to a new and detailed analysis of global temperatures by James Hansen and colleagues of NASA's Goddard Institute for Space Studies in New York, published in 2010, the average temperature of our world in the decade to 2009 was 0.8°C warmer than at the start of the 20th century. Scarily, by way of showing how quickly the planet is now heating up, two-thirds of this temperature rise has occurred since 1975. Taken as a whole, the 1990s constituted the hottest decade since the beginning of the instrumental record more than 130 years ago. This has since been superseded, however, by the period from January 2000 to December 2009, on top of which 2010 equalled 2005 as the hottest year ever.

The IPCC Fourth Assessment and beyond

In its 2007 Fourth Assessment Report, the Intergovernmental Panel on Climate Change (IPCC) presents a revealing summary of how a changing climate is already manifesting itself, or at least how it did so up until early 2006, which was the formal cut-off date for science included in the report. This is bolstered by findings since the report's publication that conspire to paint an even bleaker picture of how human-induced warming is changing our planet, particularly in the polar regions. Perhaps the key message arising from the Fourth Assessment, and more recent findings, is that the averaged-out global temperature increase figures that we read about in the papers or hear on the news hide much of the detail about what is really happening. As well as the land, for example, the oceans are warming too, and have heated up since 1961 right down to a depth of three kilometres. This is a consequence of the fact that, luckily for us, the oceans have sucked up more than three-quarters of the extra heat already added to the climate system. One result is that thermal expansion of water has made the oceans swell; as a consequence they take up more space, pushing up the rate of sea level rise and encroaching further upon the land. Elsewhere, at high latitudes, temperatures are climbing far more rapidly than across the world as a whole. In West Antarctica they have been rising by around 0.5°C a decade since the 1940s, while in the Arctic the temperature increase has been close to twice the global average in the past 100 years or so. High-latitude warming is already contributing to the rise in sea level, which has climbed from a shade under 2 mm a year between 1961 and 2003 to more than 3 mm a year in the period between 1993 and 2010. This figure may not seem likely to promote an immediate mass exodus from coastal zones, but the exceptionally rapid warming at the poles bodes ill and has huge implications for the future rate of sea level rise because, of course, this is where the vast majority of the Earth's ice resides.

Glaciers, small ice caps and snow cover across the world's mountain ranges have experienced widespread decline as warmer temperatures promote melting and reduce snowfall. However, the real action is at the bottom of our planet and, especially, at the top, where it seems that the days of the Arctic sea ice are numbered. Summer sea ice cover has been down in recent years by close to 40 per cent of its long-term average—a reduction in area equal to that of Texas and Alaska combined. At the same time, its volume has been slashed by an extraordinary 70 per cent compared to the 1979–2009 average, so the ice is getting thinner too. Both the Greenland and Antarctic ice sheets are also starting to crumble, with floating ice shelves in Antarctica breaking off, such as the Luxembourg-sized Larsen-B in 2002, and outlet glaciers draining the interiors of Greenland and Antarctica speeding up and shedding ice mass into the oceans. Latest to make an impression was north-west Greenland's Petermann Glacier, from which a chunk of ice with a surface area getting on for a quarter that of Greater London, snapped off in August 2010, the largest calving event in the region for half a century. The speeding up of Greenland's glaciers is hardly surprising given the fact that between 1979 and 2008, the area of the ice sheet over which temperatures rose above melting point during the year increased by a full 50 per cent. In the particularly warm summer of 2007, temperatures high enough for the surface ice to start melting were encountered across the whole of south Greenland. At the other end of the world the ice has been shifting faster too. Eric Rignot of the University of California, Irvine, and co-researchers, reported in 2008 that ice-sheet loss in the region

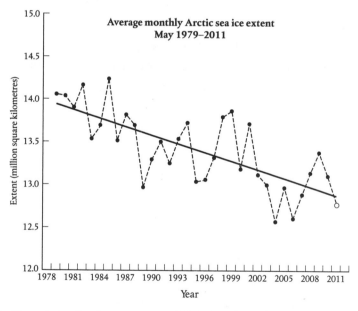

Fig 4. Recent drastic changes suggest that the days of the Arctic sea ice are numbered. Summer sea ice cover has been down in recent years by close to 40 per cent of its long-term average, while ice volume has been slashed by an extraordinary 70 per cent compared to the 1979–2009 average.

went up by nearly 60 per cent in the decade to 2006, with annual losses in the Antarctic Peninsula—the continent's long finger that points northwards towards the Falkland Islands—totalling by then sixty billion tonnes or more of ice.

One of the more obvious trends revealed in the Fourth Assessment Report and later studies reinforces the fact that, broadly speaking, wet areas are getting wetter and dry areas drier. Over the past hundred years or so, northern Europe, northern and central Asia, and much of North America have become noticeably more moist, while the Mediterranean, southern Asia, the south-west USA, and western and southern Africa have become increasingly desiccated. Big changes in the water cycle are becoming clear, with a huge and rapid increase in the volume of water pouring from the world's rivers into the oceans reported in 2010 by Tajdarul Syed of the Indian School of Mines, and his colleagues. In just 13 years, from 1994 to 2006, annual river flow rose by almost one fifth, providing hard evidence of a speeding-up of the hydrological cycle that is widely predicted in global climate models. The cause of the acceleration is mainly evaporation from the oceans driving more rainfall in places over land, but melting glaciers, thawing permafrost, and residual groundwater pumped to the surface for irrigation also make contributions. The consequences of the hydrological cycle intensifying will vary across the world, bringing more frequent flooding to some areas but drought conditions to others.

Drought has become increasingly apparent since the 1970s, particularly in the tropics and subtropics. In the south-western USA, persistent severe drought seriously threatens water supplies and has caused water levels to plunge in reservoirs such as Lake Mead—the nation's largest—which is impounded by the Great Hoover Dam. In 2010, extreme drought conditions accompanied blistering temperatures and spawned raging wildfires across Russia, severely affecting the harvest and causing an estimated 56,000 excess deaths, while in

Fig 5. Persistent severe drought in the south-west United States has caused water levels at Lake Mead – the nation's largest reservoir – to plunge. A continuation of this trend threatens water supplies to cities such as Phoenix and Las Vegas and power generation at the Great Hoover Dam, which impounds the reservoir.

2011 the Amazon region is facing its third major drought in just seven years. As I write this, severe drought in the Horn of Africa has already taken the lives of 30,000 children under the age of five and threatens 12 million people in all.

Whether heat wave, drought, storm, or flood, it is rarely possible to point a finger at a particular weather event and claim with certainty that it was the result of climate change and that it would not have occurred anyway in a world whose atmosphere and oceans were not undergoing a serious shake-up. Nevertheless, like droughts, floods have also been making themselves more conspicuous, and nowhere more so than in Pakistan in the summer of 2010, when exceptional monsoon rains led to the Indus river bursting its banks, submerging—at the peak of the floods—one fifth of the country and affecting more than 20 million people. According to the Dartmouth

Flood Observatory in the USA, the annual number of serious floods has risen rapidly from between 100 and 150 in the mid to late 1990s to close to 300 in the early years of the new millennium, and since 1990 more than 1000 river flood disasters have together taken well over 100,000 lives. Of course, flooding has always been with us and any rise in flood numbers could be the result of one or more of a whole range of factors, including natural variation in rainfall, urbanisation and deforestation, and other land-use changes. Any upward trend cannot be explained purely, therefore, in terms of a changing climate. Nevertheless, the fact that there is observational evidence for increases in extreme rainfall and river flow means that this should not be a surprise. But enough of the past; what of the future? The calm of a stable climate is clearly over and we are already experiencing the first gusts that herald the storm to come, but when will the tempest arrive and how brutal will it be?

The storm after the calm

The continued failure to act on emissions would appear to place us on a collision course with catastrophe. Signs that our world's climate is changing—and fast—are everywhere, but looking ahead exactly what can we expect as the relative climatic calm of the last few millennia gives way progressively to a tempestuous future of our own making, and how long will it be before it is obvious, even to the most entrenched denier, that our climate is being transformed? On the basis of the heat-absorbing properties of accumulating greenhouse gases, forecasting that the average temperature of our planet will increase is the easy bit, and indeed we can observe that this is already happening. Predicting how quickly and how far our planet will warm, how temperature rises will vary from one part of the planet to another, and what the implications will be in terms of melting ice sheets, rising sea levels, and changing weather patterns is quite another matter. For one thing, the nature, timing, and scale of such outcomes will depend upon human activities in coming decades; for example will we continue on a business-as-usual path that sees emissions climbing ever faster or will we come to our collective senses and adopt a more broadly sustainable lifestyle at the global level that results in emissions rising less steeply, or even falling?

To try to get some idea of what we might face, the IPCC presents a range of predictions for future temperature rises based upon a variety of hypothetical scenarios (the so-called SRES; Special Report on Emissions Scenarios) that describe how our society might develop in the course of the 21st century. Clearly producing a set of predictions based upon a clutch of (however accurately) imagined scenarios is not the way we would choose to try to forecast our planet's climate in decades and centuries to come, but it is the best we can do at the moment. The broad range of socio-economic assumptions made means that the actual increase in global average temperature experienced by the end of the century is likely to be somewhere within a wide array of output temperatures, spanning 1.1°C to 6.4°C, regurgitated by the climate models. The Fourth Assessment Report does narrow this range somewhat by suggesting best estimates for the different scenarios, thereby projecting a most likely rise by 2100 of between 1.8°C and 4.0°C. A similar range of estimates is presented for future sea levels which the IPCC assessment predicts will be between 18 and 59 cm higher by the century's end. Other projections pretty much involve more of the same; in other words, a confirmation and continuation of the trends that we are already experiencing. So this means the continued retreat of snow and ice cover across the planet, with the eventual loss of Arctic sea ice altogether during late summer; the expansion and deepening of the permafrost thaw zone; more frequent extreme temperatures, heat waves, and intense precipitation events; and very probably windier and wetter tropical cyclones. High latitudes are very likely to experience increased precipitation, while reduced rainfall at lower latitudes will probably foster the potential for more droughts in the subtropics. The assessment even predicts a slow-down in the Meridional Overturning Circulation (MOC), the system of ocean currents, including the Gulf Stream, that bathes the UK and northern Europe in relatively warm water from the tropics. Sudden shut-down of the circulation, however, is held to be very unlikely, while a consequent return to full ice age conditions, as portrayed in the Hollywood blockbuster, *The Day After Tomorrow*, remains rooted firmly in the realms of science fantasy.

Bleak and bleaker

To some, the picture painted of the Earth's future in the IPCC Fourth Assessment is scary; to those who deny that climate change is actually happening or that it is a consequence of human activities, it is a travesty. To still others, it is over-optimistic and too conservative. The latter accusation is understandable, given that prior to publication every sentence is scrutinized to ensure acceptability by representatives of the world's governments, many of whom can hardly be described as sympathetic to the idea of anthropogenic climate change; the USA and Saudi Arabia spring particularly to mind. The charge of optimism is justified too, and the report's authors recognize that they have to be pragmatic in order to get any worthwhile consensus agreement on content. They acknowledge, therefore, that neither the potential break-up of the polar ice sheets nor possible feedback effects, such as massive methane release due to wholesale permafrost thawing, are adequately addressed in the Fourth Assessment Report. The explanation provided is that the science in respect of these and other mechanisms is not yet fully understood or is poorly constrained. The hard-nosed perspective, however, has to be that an assessment that explicitly flagged up the potential ramifications of ice sheet collapse, massive methane exhalation across the Siberian tundra, and the failure of the oceans to any longer act as an effective sponge for our carbon pollution and heat, would have stood very little chance of seeing the light of day.

Nonetheless, since publication of the Fourth Assessment, new research has tightened up much of the science in relation to what were once regarded as worst-case scenarios, but which are now entering the mainstream. The portfolio of ramifications that anthropogenic warming may bring is also widening; the potentially hazardous response of the Earth's crust, which forms the theme of this book, for example, is addressed in the recent IPCC report on climate change and extreme events. At the same time, some climate scientists unrestricted by the politics of the IPCC framework are becoming far more vocal in expressing their views about the nature and pace of future climate change. This growing trend for climate scientists, who have been criticized for keeping too low a profile, to tell it as they see it, is beginning to ensure that governments and the public are at last being given the opportunity to appreciate what unmitigated climate change may really mean for future society. A reflection of the new mood can be found in the synthesis report of the Climate Change Congress held in March 2009, a coming together in Copenhagen of 2500 scientists and others, nine months prior to the disastrous UN COP15 Climate Change Conference that failed spectacularly, in the same city, to deliver any meaningful deal on greenhouse gas emissions reductions. Based upon post-IPCC Fourth Assessment knowledge, the synthesis report makes chilling reading. Alarmingly, its authors draw attention to the fact that observations since 2007 reveal that some key climate indicators are changing close to the upper end of the range of IPCC projections and, in the case of sea level, far outstripping them. With the oceans now known to have warmed by half as much again as reported by the IPCC, and with accelerating melting of polar ice sheets contributing progressively more towards ocean volumes, global sea level is now expected to be one metre or more higher by 2100, almost double the worst case presented in the IPCC Fourth Assessment. Neither will sea level cease to climb at the end of the century. However successful we are at reducing emissions, it will continue to escalate for centuries thereafter as thermal expansion of the slowly warming oceans continues and the melting and crumbling ice sheets in Greenland and Antarctica persist in contributing to the rising water levels. The synthesis report also flags up the longevity of anthropogenic climate change, reminding us that the consequences of our activities today will still be felt 50 generations and more down the line. Because of the enormous inertia in the climate system, even when (and if) anthropogenic greenhouse gas emissions are reduced to nothing, global temperatures will barely fall at all for at least a thousand years.

Increasing desperation, as the global community apparently sleep-walks towards climate catastrophe, is providing the impetus for some climate scientists at least finally to speak out as individuals. This is reflected in the August 2010 testimony of Richard Alley to the US House of Representatives, during which the Pennsylvania State University climate scientist pulled no punches. Alley warned that a critical tipping point could be exceeded within the next 10 years that could mean the ultimate loss of the Greenland Ice Sheet, committing the world to an eventual sea level rise of up to seven metres, and expressed the opinion that 'what is going on in the Arctic now is the biggest and fastest thing that Nature has ever done'. The bleak prospect for any effective action on emissions has also led to James Hansen speaking out even more volubly than usual. In his call-to-arms book, *Storms of my Grandchildren*, arguably the world's best known climate scientist warns unapologetically of what might be the ultimate legacy of our society's unthinking behaviour—a broken and baking, ice-free world with sea levels 70 metres or so higher than they are today. Hansen's exasperation at the absence of effective action to bring down emissions is all too apparent when he compares the current situation to that faced by Abraham Lincoln over slavery, and Winston Churchill over the rise of Nazi Germany, Hansen's despairing call that 'the time for compromise and appeasement is over' echoes a frustration shared the world over by climate scientists who may be less able or less prepared to speak out, and by countless others who dread what a carbon-soaked future will mean for their descendants.

As to the precise form this future will take, even with the best will in the world modelling something as complex and interactive as the Earth's climate may simply never be able to provide us with an accurate picture of what our planet will look like 50 or 100 years from now and beyond. As physicist Niels Bohr first and most famously observed, and many others have since reiterated, 'predicting is very difficult; especially about the future'. Maybe then, the best way to gauge the nature of the world to come is to look back rather than to project forward. Zeroing in on the post-glacial period, for example, reveals the many and varied ways in which a dramatically changing climate evoked a dynamic response from the crust beneath our feet, and might do so again. Delving further back in time, studies of previous occasions when carbon-dioxide-enriched atmospheres were the order of the day, may leave us far more enlightened about the conditions that increased greenhouse gas concentrations might bring than any number of computer projections.

Once and
Future Climate

A common failing of human beings, both as individuals and collectively, is their apparent inability to imagine a world or a life any different from the one they are experiencing at any given time. Arguably, this is the greatest obstacle faced by climate scientists trying to make national governments, institutions, businesses, and people generally wake up and pay attention to the massive threat they face—to appreciate what an appalling impact unmitigated anthropogenic climate change is slated to have on our world and our descendents. Maybe there is some sort of in-built safety mechanism, a sort of comfort-blanket, which makes us assume that tomorrow will be pretty much the same as today; that the world when we are middle-aged or old will be, broadly speaking, the same as it was when we were young. In recent centuries and in the fastest-developing nations, however, this has never been the case. Either by evolution or by revolution, the world to which we bid farewell at life's end is almost always transformed from the one into which we were born. Think, for example, of the many septuagenarians whose birth cries intruded on the prim, proper, and pomp-driven world of Victorian England, who lived to see—if not make the most of—the 1967 Summer of Love, and the first manned landing on the Moon two years later.

Whatever our ingrained comfort blankets assure us about continuity and the status quo, nothing ever stays the same, and our climate provides a prime example. It is barely a century since much of the northern hemisphere shivered in the bitter winters of the so-called Little Ice Age, during which a fractionally dimmed Sun fostered a cooler climate that persisted, with some short respites, from the 17th century to the end of the 19th. Now, our communal behaviour, augmented by a widespread denial of the possibility of change, is pushing us rapidly in the opposite direction, bringing us to the brink of hothouse conditions that, as things stand, are virtually certain to transform our planet and our society, making tomorrow's world very different from today's. Look back over a much longer time scale, and far greater variations in our climate are recognized. While these are a reflection of natural changes in the environment of our planet, rather than consequence of human activities, some past climates can provide a useful, and somewhat terrifying, guides to what our planet might look like by 2100 and in the centuries that follow. In particular, the post-glacial period provides us with the perfect opportunity to examine and appraise how abrupt and rapid climate changes drive the responses of the solid Earth, which form the focus of this book. Peering further back in time, to warmer episodes with more carbon-enriched atmospheres, supplies us with important clues about how much hotter our world might become and how far sea levels could rise. This, in turn, can help us evaluate the ultimate scale and extent of anthropogenic climate change, thus allowing us better to weigh up the chances of a future hazardous response from the Earth's crust.

It is more than two centuries since Scottish naturalist, James Hutton, laid down the principles of so-called uniformitarianism, a key tenet of the Earth sciences that assumes that the natural processes and mechanisms that we see happening today have always been the same and can therefore be used to explain what we observe in the geological record. This philosophy is succinctly portrayed by the phrase 'the present is the key to the past'. Earth scientists are also recognising that we can learn much about our contemporary world by looking to earlier episodes in Earth history. Furthermore, and in the context of climate change in particular, there is much we can learn about what our world might look like in 2100 and beyond by subjecting to detailed scrutiny periods in our planet's past when conditions were comparable to today's. If care is observed in selecting appropriate analogues and note is taken of various caveats, the past can—where the Earth's climate is concerned—provide a useful guide to the future. Nowhere is this better demonstrated than during the Cenozoic, our planet's most recent era and perhaps, from a climate point of view, its most dynamic.

Introducing the Cenozoic

Human's love categorising and pigeonholing, and the evidence for this is as compelling in the way we have divided up and labelled the 4.6-billion-year history of our planet as it is in the apparent need that many (mostly men it seems) feel for alphabetising and cataloguing CD collections. For example, since the end, of the Pre-Cambrian around 540 million years ago,—the immense span of time making up about the first seven-eighths of our planet's history—geological time has been divided into three eras: the Palaeozoic, the Mesozoic, and the Cenozoic, each itself comprising a number of geological periods, such as the Devonian, Triassic, and Cretaceous, to name just three out of a round dozen. From a climate viewpoint, the Cenozoic— meaning 'new life', from the Greek *kainos* 'new' and *zoe* 'life'—which is sometimes also referred to as the Tertiary, is perhaps the most fascinating period; it is certainly the most relevant in terms of providing a mirror within which visions of our future world may be glimpsed. The Cenozoic started with a bang, following on immediately after the asteroid impact, voluminous outpourings of lava—or a combination of the two—that resulted in the mass culling of the dinosaurs and countless other species 65 million years ago. The Cretaceous–Tertiary (K–T) or, more correctly, Cretaceous–Palaeogene (K–Pg) mass extinction literally marked 'the end of an era', defining the boundary between the Cretaceous, the last period of the Mesozoic Era, and the Palaeogene, the first period of the Cenozoic.

The Cenozoic climate is a real roller coaster, starting with sweltering conditions and gradually but inexorably cooling off, so that for the past couple of millions years our world has frequently been held firmly in the grip of ice. Today, in contrast, it is playing host to an episode of very rapid warming, and this not for the first time—as will soon become apparent. However rapidly the Arctic sea ice is currently dwindling, the waters of the Arctic Ocean remain bitterly cold and are prone to freezing at the drop of a hat. It is hard, therefore, to imagine a time, not too far back in the geological past, when these frigid waters played host to cruising crocodiles, turtles, and other species whose natural habitats today lie several thousand kilometres further south. Fantastic-sounding but true nevertheless. The Palaeocene—the earliest Epoch of the Palaeogene—was a sweltering world with ice-free poles and palm trees growing in Russia's

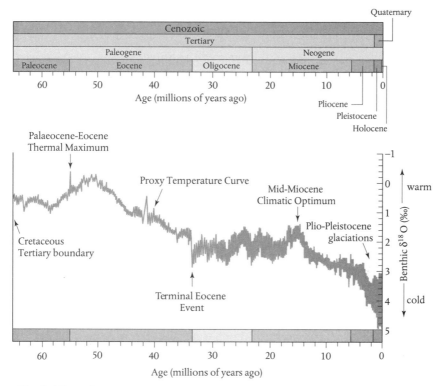

Fig 6. The roller-coaster climate of the Cenozoic Era was one of great contrasts, but the broad trend was towards a cooling of the planet. After temperatures peaked in the hothouse world of the Palaeocene-Eocene Thermal Maximum (PETM), and notwithstanding the odd respite, it was pretty much downhill all the way.

Kamchatka Peninsula. As far as evolution was concerned, it was a time of innovation, with new groups of organisms arriving on the scene to occupy the niches vacated during the K–T mass extinction. Indeed, the name of the epoch, from the Greek, refers to the 'older (palaios)—new (kainos)' transitional faunas of this time, before the emergence of modern mammals during the succeeding Eocene Epoch. The concentration of carbon dioxide in the Palaeocene atmosphere is not well established and it could well have been more than 1000 ppm or even significantly less. What is known, however, is that global average temperatures were as much as 15°C higher than they are today. Fossil finds provide evidence for subtropical vegetation growing in Greenland and in Patagonia in the southern hemisphere, while the ice-free poles in this clement Palaeocene world were characterized by climates that were cool rather than frigid, and covered by forest rather than the bleak and frozen tundra that stretches across most high latitude regions today. Towards the equator, warm and humid conditions encouraged the growth of widespread tropical and subtropical forests that probably looked little different from the rainforests of today, or at least the bits that have not yet been grubbed up or torched. To those who favour constant warmth above bracing changes in seasonal temperature, the Palaeocene would have been blissful. And it was to get hotter still—far hotter, as the planet's climate underwent an astonishing change, the like of which has been seen on very few occasions in the Earth's long history.

The great heat spike mystery

A warming path already established in the late Palaeocene saw a rising trend in temperature that was steady and nothing really to write home about. Suddenly, however, a little under 56 million years ago, and for reasons that are still not fully understood, the average global temperature shot up by 6°C over a time span as short as 10,000 years, with the poles heating up by 10–20°C over the same period and the surfaces of the oceans warming by up to 9°C in less than 10 millennia. In the Antarctic, sea-surface temperatures reached 20°C, with Arctic Sea temperatures climbing to a high of 24°C, in both cases substantially warmer than the rather meagre summer sea temperatures off the invigorating British east coast resort of Skegness. The sudden warming had a dramatic effect on ecosystems, most notably involving the emergence on land of a new mammal fauna that included the primates and horses, and a mass extinction of the tiny, single-celled, marine organisms known as foraminifera, that lived deep in the oceans. This so-called 'transient warming event', which lasted for as long as 170,000 years, is the most prominent in the whole of the Cenozoic Era. It also marks the transition from the Palaeocene Epoch to the Eocene and so goes by the name of the Palaeocene–Eocene Thermal Maximum, or simply, the PETM. Such sudden outbursts of heat are rare but not unique, and a few others are recognized further back in time, during the Cretaceous, Jurassic, and Triassic periods. As in the case of current warming, the ultimate cause of these warm 'spikes' in the global temperature record seems to be a sudden rise in the concentration of atmospheric carbon dioxide. The start of the PETM, for example, coincides with the addition of a huge mass of the gas to the atmosphere, estimated to contain between about 4000 and 7000 billion tonnes of carbon. For comparison, human activities in 2009 released greenhouse gases equivalent to about 8.3 billion tonnes, a drop in the ocean in comparison, although a drop that is repeated year after year. The great enigma of the PETM relates to the source of this vast quantity of carbon dioxide, and lively debate has for some time addressed where it could have come from and how it might have found its way into the atmosphere. A long list of candidate causes includes a comet impact charged with importing the extra carbon into the Earth system from beyond the atmosphere, although this has very few supporters. A second proposes the ignition and burning of vast areas of peatland, which seem to have been particularly abundant at the time. As more than 90 per cent of the world's biomass would need to be reduced to ashes to match the scale of the extra carbon dioxide in the atmosphere, this suggestion too is not a popular one. An increase in volcanic activity has also been put forward, but again the amount of additional carbon dioxide pumped out during even the most intense hike in volcanic action cannot match the required level. Nevertheless, according to one model at least, rising magma may still have had an important role to play, about which more later.

The favoured carbon-source candidate is, far and away, the wholesale release of methane from so-called gas hydrate or clathrate deposits. These are solid, ice-like, mixtures of water and gas—usually methane—stored in marine sediments and trapped in Arctic permafrost in truly prodigious quantities. Built from 'guest' gas molecules enclosed within cage-like structures built from water molecules, gas hydrates are stable at low temperatures and relatively high pressures. They can, however, become destabilized by warming or pressure reduction, resulting in dissociation and release of the methane in its gaseous state. Methane contains carbon and is also a very effective greenhouse gas in its own right, making it understandable that worries over how gas hydrates might be affected by anthropogenic warming have been growing. Bearing in mind that every cubic metre of gas hydrate can produce 163 times as much methane, it is also hardly surprising that the world's energy companies have been licking their lips at the prospect of getting their hands on these somewhat esoteric hydrocarbon deposits at a time when conventional supplies are dwindling. This confluence of concern and covetousness has driven a burgeoning programme of research so that far more is now known about these rather enigmatic deposits than a couple of decades ago. Notwithstanding this, it has taken some time to pin down just how much gas hydrate is out there. Usually expressed in terms of their contained carbon, initial estimates were in excess of 10,000 billion tonnes, an order of magnitude higher than the total carbon held in the atmosphere in the form of carbon dioxide and methane, and far more than in all other fossil-fuel sources combined.

Since these early approximations, global gas hydrate resources have been revised down considerably to something closer to 2000 billion tonnes of carbon. This is smaller than the 5000 billion tonnes of carbon locked away in other fossil fuel sources, but still about two and half times the amount of carbon currently held in the atmosphere. Hence the continued concern of climate scientists and the enduring interest of the energy companies.

Returning to the PETM, the evidence for a gas hydrate role in the sudden warming is strong. One strand of support comes from giant, submarine, sediment slides along the margins of the continents, particularly in the western Atlantic, hinting at the sort of widespread destabilisation expected as the solid hydrate breaks down to much greater volumes of methane gas. A second comes from the distinctive carbon signature associated with methane derived from gas hydrates.

Like many elements, carbon comes in a variety of forms, or isotopes, each of which has an atomic structure containing slightly different numbers of neutrons. By far the most common isotope of carbon (99 per cent of all naturally occurring carbon) contains six protons and six neutrons, and is known as carbon-12 or C^{12}. About one per cent of carbon comes in the form of isotope carbon-13 (C^{13}), which contains an extra neutron, while just a trace of natural carbon has two extra neutrons, making the isotope carbon-14 (C^{14}). Both carbon-12 and carbon-13 are stable isotopes, but carbon-14 is radioactive, ultimately breaking down to form an isotope of nitrogen. This property of carbon-14 forms the basis of the radiocarbon dating method, which has wide application in archaeology and geology. As the total number of protons and neutrons determines the mass of an element, carbon-12 is the lighter of the two stable carbon isotopes. The carbon held in methane contained in gas hydrates is much depleted in the heavier carbon-13 relative to the lighter carbon-12 isotope, so that its carbon 'signature' is enriched in the lighter isotope. On this basis, the addition of very large volumes of hydrate-sourced carbon to the PETM environment can be detected by analysing the carbon isotope composition of marine sediments deposited at the time. Such analysis reveals the existence of what is technically known as a Carbon Isotope Excursion (CIE) in ocean waters, which is another way of saying that their isotopic composition underwent a sudden and significant change with respect to the relative proportions of the three carbon isotopes. The CIE is described as negative where it involves enrichment in the lighter carbon-12 isotope. In the case of the PETM, a negative CIE is identified amounting to about four per cent, indicating the addition to the late Palaeocene oceans and atmosphere of a significant volume of carbon depleted in carbon-13 relative to carbon-12.

While the sudden change in the carbon isotope composition provides a convincing 'smoking gun' in relation to a gas hydrate role in the global temperature spike that marked the PETM, a couple of key questions remain to be answered before this can be accepted as the whole story. Were gas hydrates responsible, on their own, for releasing all the carbon that initiated the sudden warming at the end of the Palaeocene, and why and how was the wholesale breakdown of gas hydrates triggered? Current thinking suggests that there was almost certainly not sufficient gas hydrate around at the time of the PETM to explain all the additional carbon. As mentioned earlier, today's inventory of gas hydrate is unlikely to provide more than 2000 billion tonnes of carbon and in the warming Palaeocene world, gas hydrate reserves are likely to have been even smaller. If this was the case, then the rest—probably the bulk—of the extra carbon must have come from somewhere else. This hints at a situation wherein rapid warming, driven by carbon dioxide accumulating in the atmosphere from some unknown source, heats up the ocean waters to such a degree that wholesale dissociation of gas hydrate deposits is triggered in a positive feedback effect that releases large volumes of methane, warming the planet even further. Within this scenario, while gas hydrates contribute to warming at a later stage, their breakdown is a response to initial warming due to an event or events unknown. A recent study of the environmental conditions in the very late Palaeocene by Appy Sluijs of Utrecht University in the Netherlands, and his co-researchers, suggests that this is just what happened. Examination and analysis of the remains of particular single-celled organisms contained within marine sediments enabled the temperature of the ocean at this time to be determined, revealing that the planet started to warm significantly several thousand years prior to the CIE arising from gas hydrate breakdown. Unfortunately, the work of Sluijs and his team does not shed any light on the enigmatic source of the carbon that drove the initial warming. Others, however, have pursued the quest to pin down its origin, which has led them to the North Atlantic and to the dramatic geological events that were unfolding there during the late Palaeocene.

Prior to around 200 million years ago, the Atlantic Ocean did not exist. In the early part of the Jurassic Period, however, the ancient supercontinent of Pangaea started to fragment, opening a narrow proto-North Atlantic Ocean that separated the new northern supercontinent of Laurasia from its southern counterpart, Gondwana. By Palaeocene times, more than 100 million years later, the Atlantic was a wide ocean that extended southwards, separating South America and much of North America, in the west, from Africa to the east. Coincident with the PETM, the North Atlantic was undergoing its final extension, slicing northwards between Greenland and northern Europe. As might be expected, the splitting of a continent tends to involve quite a kerfuffle, and this was no exception. Accompanying the event, wholesale melting in the Earth's mantle underlying the region fed vast outpourings of lava across Canada's Baffin Island, Greenland, the Faeroes, and north-west Britain. In places, lavas were piled more than seven kilometres thick, while elsewhere magma intruded en masse into the local rock and sediments. Estimates suggest that the total volume of magma involved was staggering, ranging between 5 and 10 million cubic kilometres. Impressive, undoubtedly, but what has this to do with the PETM? According to Mike Storey of Roskilde University in Denmark, and colleagues, quite a lot. Storey and his fellow researchers propose that the puzzle of the initial PETM warming can be explained by the release of prodigious volumes of carbon-12 enriched methane as magma associated with the splitting of Greenland from Europe heated and baked carbon-rich sediments that floored much of the region prior to the tectonic upheaval. The timing is just about right, with the start of the PETM occurring a little after the beginning of the great, magmatic outburst. This link remains a hypothetical one and there is, as yet, no definitive evidence that connects the two events in a cause-and-effect manner. Nevertheless, the idea provides, at the very least, a useful stop-gap while other possibilities are investigated.

The PETM revisited?

Interest in the world of the late Palaeocene–early Eocene was sparked and sustained by the possibility that what happened at the PETM might provide us with a glimpse of how our world might end up should we fail to reduce greenhouse gas emissions rapidly and sufficiently enough. Most critically, this unusual event provides us with a laboratory within which to examine the effects and consequences of the wholesale release of carbon into our planet's atmosphere over a very short period of time. The bottom line is that the hothouse world of the PETM may mirror conditions on the planet in the coming century and beyond. To revisit some salient points, the PETM warming appears to have resulted from the release of 4000–7000 billion tonnes of carbon over a period of maybe 10,000 years, resulting in a global average temperature rise of about 6°C. It is sobering to reflect that human activities have generated around 500 billion tonnes of carbon, and continue to do so at a rate of more than eight billion tonnes a year. Even in the depths of the biggest recession since World War II, human emissions in 2009 rose by 1.3 per cent. In 2010, as the global economy perked up, this figure shot up by a staggering 6 per cent compared to the previous record year in 2008, with no prospect of a significant slow-down in sight, let alone any reduction.

In just a couple of centuries then, we have released up to one eighth of the amount of carbon that drove the PETM. Given the current rate of deforestation and the fact that conventional fuel reserves are estimated to provide around 5000 billion tonnes, it would not be at all unreasonable to imagine that eventually the carbon released by our activities will tally with that which triggered the PETM. The especially scary thing is that this would likely be accomplished in a fraction of the time. In addition, it may well be that we don't need to match, through our own activities, the carbon release at the PETM to trigger a sudden, rapid, and persistent rise in global temperatures. As the world continues to warm in response to anthropogenic emissions, so the conditions are predicted to become less favourable for life, the land, and the oceans to take up their share of the excess carbon. In another example of a positive feedback effect, a warmer world is likely to mean that these 'carbon sinks' suck up progressively less carbon, as a consequence leading to its more rapid accumulation in the atmosphere and even more swiftly rising temperatures. It is even possible that some carbon sinks may become carbon sources, reversing their current roles so as to add carbon to the atmosphere rather than subtract it. This is particularly worrying, bearing in mind the amount of carbon currently held in sinks: more than 2000 billion tonnes in plants and soils and a massive 40,000 billion tonnes in the waters of the world's oceans and the organisms that live within them.

More disturbing still is the fact that evidence is already coming to light that some carbon sinks are showing signs of becoming less effective. The oceans, for example, have—fortunately for us—already absorbed somewhere between one fifth and around one third of all the carbon produced by human activities, but they may not be able to continue to provide this service for ever. According to Samar Khatiwala at Columbia University's Lamont-Doherty Earth Observatory, and associates, the uptake of anthropogenic carbon by the oceans rose sharply after 1950, but the rate of uptake has started to decline in recent decades. The worry is that the sheer volume of carbon we are producing is showing signs of overwhelming the oceanic carbon sink, perhaps as warmer waters—that are also made more acidic as they absorb more carbon—find it more difficult to take up as much carbon as they used to. Should this trend continue, then more of the carbon we produce will stay in the atmosphere rather than becoming locked up in the oceans, so enhancing greenhouse warming.

Building estimates of carbon feedback effects into model projections of future climate makes for sobering forecasts of temperature rises in the decades ahead. In a dedicated issue of the Royal Society's Philosophical Transactions (A), published in late 2010, Richard Betts of the Met Office Hadley Research Centre and colleagues summarize how quickly, taking into account carbon-cycle feedback effects, the world might have to face a 4°C average temperature rise; the results are both astonishing and alarming. Assuming the pessimistic and fossil-fuel dominated SRES scenario, A1FI, the choice justified by the current rate at which emissions are climbing, Betts and co-researchers provide a best estimate that suggests that global average temperatures may reach 4°C as soon as the 2070s. Even worse, if positive feedback effects are particularly strong, this figure could be reached as soon as the early 2060s. Think about this for a moment; that is a temperature rise twice that widely charged with equating to a guardrail that must not be crossed if the most serious effects of climate change are to be avoided, arriving in less than 50 years. We are not talking here of centuries down the line, in the dim and distant future of our far-removed descendents, but of a time when my children—and perhaps yours too—will still be in the prime of life. Of course, it is possible that we will collectively come to our senses in the near future and initiate plans designed to make serious inroads into global emissions. From the perspective of August 2011, however, the prospects for such a change in thinking seem very poor. Consequently, it is perfectly possible that not long after mid-century our world will be well on its way to the hothouse Earth conditions of the Eocene, having passed the 2°C dangerous climate threshold as early as 2030 or soon after. The upshot could be a 5–6°C temperature hike, comparable to that of the PETM, by the end of this century. As it seems to have done during Palaeocene times, there is also the possibility that this initial anthropogenic warming could ultimately lead to the wholesale dissociation of gas deposits as the warming thaws Arctic permafrost and penetrates to ever greater depths in the oceans, leading to even greater warming. The fact that large volumes of methane are already venting into the atmosphere from methane-rich sediments beneath the shallow seas off the coast of Arctic Siberia cannot be anything other than bad news. If we do manage to conjure up another PETM-like event we should not expect temperatures to return to what we have long regarded as normal values any time soon. Remembering that the PETM lasted for perhaps 170,000 years, we would need to face up to the fact that it would likely be a hundred millennia or more before temperatures returned to 20th century values.

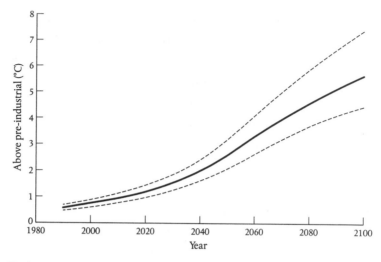

Fig 7. Assuming that emissions rise in line with the pessimistic and fossil-fuel dominated SRES scenario, A1FI, a best estimate suggests that the global average temperature could reach 4°C as soon as the 2070s (see thick central line, which shows the medium projection). Even worse, if positive feedback effects are particularly strong, this figure could be reached as soon as the early 2060s (see the upper dotted line).

Cool with warmer intervals

Even once the PETM waned, temperatures on Planet Earth remained far higher than they are today and the following five or six million years—a time known as the Early Eocene Climatic Optimum (EECO)—remained balmy and humid. After this, things started to go downhill as the whole character of the Cenozoic Era began to change. Global temperatures fell throughout the rest of the Eocene Epoch, and the Earth's climate became broadly cooler and drier. This was the start of a long road that would ultimately lead from the hothouse to the icehouse. The steady drop in temperatures was accompanied by relentlessly falling levels of atmospheric carbon dioxide that by the middle of the succeeding Oligocene Epoch were down to around

500 ppm. The long, steady, temperature fall throughout the Eocene was interrupted at the start of the Oligocene, some 34 million years ago, by a dramatic cooling known as Oi-1, which marked a major transition in the planet's climate. The temperature of deep ocean waters fell to below 3°C, down from 12°C during the warm conditions of the early Eocene, and large ice sheets started to form for the first time in Antarctica, leading to sea levels plummeting by 55 metres. In response to the rapid cooling, vegetation zones underwent a major transformation, with tundra and steppe replacing northern forests and the shrinking of tropical broadleaved forest. Various hypotheses have been proposed to explain why ice sheets started to build on Antarctica and analysis of their relative merits is outside the scope of this book. A widely held view, however, links conditions suitable for ice accumulation with the separation of the Antarctic continent from South America and Australia. This isolated the continent, leaving it surrounded by ocean, and paving the way for the formation of the cold Antarctic Circumpolar Current, which acted as a barrier to warmer waters from lower latitudes. This is suggested to have led first to the widespread accumulation of sea ice, and later to the progressive build-up of ice sheets on land.

The Cenozoic climate roller coaster switched direction suddenly once again at the very end of the Oligocene Epoch around 25 million years ago. This time a rapid rise in global temperatures saw the Antarctic ice retreat and sea levels shoot up once again. As the Oligocene was succeeded by the Miocene Epoch (the first epoch of the Neogene Period) a little under 24 million years ago, another change of direction led to a brief cooling episode, known as Mi-1, which saw the Antarctic ice expand again and sea levels drop suddenly. This short-lived cold snap soon gave way to warmer and more humid conditions that persisted for the first half of the Miocene Epoch. The coniferous forests that had succumbed to the tundra during

Oligocene times made a comeback, and between 17 and 15 million years ago, during the so-called Mid-Miocene Climatic Optimum (MMCO) temperatures at mid latitudes were 3–6°C higher than today. Like the PETM, the MMCO may provide us with some useful clues about how anthropogenic climate change might develop in the decades ahead. As mentioned in the previous chapter, the MMCO was the last time that carbon dioxide levels in the atmosphere were higher than they are now. According to the University of California's Aradhna Tripati, and her co-researchers, concentrations were only in the region of 400–450 ppm, the sort of values we might expect within a decade if anthropogenic emissions continue to rise as they are doing. Assuming that atmospheric carbon dioxide levels and global average temperature are linked today in a comparable manner to how they were during the MMCO, then we can expect atmospheric carbon dioxide levels only slightly higher than they are now to eventually drive the global average temperature up by 3–6°C compared to pre-industrial times. This is substantially higher than the 2° rise that is often equated with keeping the concentration of atmospheric carbon dioxide below 450 ppm. These higher global average temperatures would almost certainly mean double-figure rises at high latitudes and wholesale melting of the Greenland and West Antarctic ice sheets. Such a prospect is reinforced by the observations of Tripati and colleagues that during the MMCO there was little land ice or sea ice in the Arctic, nor were there floating ice shelves in Antarctica. In fact, neither an extensive cover of Arctic sea ice nor a large permanent ice sheet on East Antarctica were able to form until carbon dioxide levels in the atmosphere fell below the 350–400 ppm threshold, during the period of rapid cooling and glacial expansion that followed the MMCO, while large ice sheets were not able to form on Greenland or West Antarctica until carbon dioxide levels in the atmosphere had dropped even further, to below 300 ppm.

To summarize, what the MMCO seems to be telling us is that atmospheric carbon dioxide levels just a little higher than they are now, and which we are virtually certain to achieve in the next few decades, will—if maintained for long enough—result in a global average temperature rise towards the top of the range projected in the IPCC's Fourth Assessment report. Wholesale melting of polar ice is likely, perhaps resulting in sea levels that eventually match those of the MMCO, which were between 25 and 40 m higher than today. The warm world of the MMCO was not to last, and a sudden cooling of 6–7° around 14 million years ago heralded a significant build-up of ice at both poles. As the Miocene Epoch drew to a close a little over five million years ago, ice covered both Greenland and Antarctica, and while warm interludes were still to come, the transition to icehouse Earth was now well and truly underway.

The Pliocene: another glimpse of our future world?

As the Miocene Epoch gave way, 5.3 million years ago, to the Pliocene—the second of the two epochs making up the Neogene Period of the Cenozoic Era—the planet continued to cool, if at a somewhat reduced rate. The climate was still warm compared to today, and during the Middle Pliocene the Cenozoic climate switch-back reversed direction again so that around four million years ago global temperatures started to climb once more. Just over three million years before present, Pliocene temperatures peaked during the 300,000-year-long Mid-Pliocene Warm Period (MPWP), during which global average temperatures were about 2–3° higher than they are today. In common with the MMCO, this relatively warm episode during the Pliocene can be looked to for clues as to where contemporary climate change may eventually take us. The warmer climate of the MPWP saw conifer forests replace much of the tundra that had become established during preceding cooler times and even establish themselves in northern Greenland. As we may see again in as little as a few decades, the Arctic sea ice cover vanished entirely during the summer months, while the Antarctic ice shelves were much reduced in size. Possible causes of the MPWP are manifold, but atmospheric carbon dioxide levels close to those that prevail at present are widely regarded as the major—or at least one major—driver of the higher temperatures. As such, the Mid-Pliocene Warm Period is held by many scholars of past climates as representing an equilibrium climate response to atmospheric carbon dioxide levels similar to those of today. Roughly translated, this suggests strongly that if we maintained these levels of the gas over a long period, our climate would approach and ultimately match that of the MPWP. While this would likely not mean global average temperatures as high as during the MMCO, it would lead to a world warmed at least to the 2°C dangerous climate-change 'guardrail'. Climate models invariably tell us that this will mean less ice at the poles and higher sea levels, and this is also the message from the warmest Pliocene times. Over the entire third of a million years or so of the MPWP, high latitude ice volumes and sea levels seem to have gone up and down like yo-yos, but at the peak of the warmth they were 30 m higher than at present, sufficient today to leave every one of the world's coastal towns and cities well and truly submerged.

After the drawing to a close of the Mid-Pliocene Warm Period the climate deteriorated rapidly and it was pretty much downhill all the way for both global temperatures and atmospheric carbon dioxide levels. By the end of the Pliocene, carbon dioxide concentrations were down at 200 ppm or below, Greenland was ice covered, and, in the northern hemisphere, the ice was moving southwards to mid latitudes. Over a period of just a few hundred thousand years, ice spread across much of the European and Asian Arctic, across Alaska and into north-eastern North America. At last, the general global cooling trend that had started in the Eocene nearly 50 million years previously had become sufficient to trigger a true ice age, the first since the glaciers ventured out of their polar fastnesses close to 250 million years earlier, during the Carboniferous Period. The switch from temperate world to ice world took place rapidly, at least in geological terms, but the various elements needed to trigger a major ice age had been slotting into place for some time. Only when all were in position could the ice accumulate sufficiently and reach out from the poles.

Making an ice age

Kick-starting an ice age is a bit like growing a rare and delicate orchid. The initial conditions have to be perfect: in the former case, the geology, atmosphere and oceans have to be set up just so, in order to allow sufficient cooling to take place so that ice starts to form at the poles. This in itself, however, may not be enough, and ice ages have withered and died without getting any further, as happened during the Oligocene and Miocene epochs. Careful nurturing is required to promote the expansion of the ice sheets beyond their polar homelands. The fact that ice ages are uncommon in Earth history suggests that they are products of conspiracies of circumstances that do not often arise. In other words, there is a recipe for an ice age, the ingredients of which are only rarely brought together. One of the key requirements is for the continents to be so arranged as to seriously limit or prevent the flow of warm water from the tropics to the poles, thereby fostering cooler conditions at the top and bottom of the world. When the polar regions are not land-locked and ocean waters at high latitudes are free to mix with warmer waters from the tropics, the temperature gradient from equator to pole is reduced to as little as 30°C. Under these conditions, sea ice still forms, both north and south but is not persistent and does not accumulate as average temperatures at the poles are not permanently below freezing. In contrast, when the polar regions are occupied by a landmass or are enclosed by land, mixing of the colder high latitude ocean waters with warmer water from the tropics is dramatically reduced, resulting in a much larger temperature gradient from equator to pole. This is the situation that prevails today, with a gradient from the Earth's midriff to its southern extremity of 65°C, and to its northernmost point, of 50°C. In the south, the existence of the Antarctic continent means that there are no polar waters to mix with those from warmer climes. Furthermore, the encircling Antarctic Circumpolar Current acts to cool the continent further. In the manner of a refrigerator, this continuously circulating current constantly sucks heat from the continent and discharges it northwards into the warmer waters of the Atlantic, Pacific, and Indian oceans. The Antarctic continent also provides the solid base upon which vast ice sheets can and have accumulated. At the top of the world, North America and Siberia form an almost complete encircling barrier that contains the cooler polar waters and the sea ice that forms from them. Today, then, geographical conditions are just about perfectly suited to maintaining ice age conditions, current anthropogenic warming notwithstanding of course. On their own, however, they do not seem to have been sufficient to trigger an ice age. After all, the disposition of the landmasses at high latitudes has been little different for the past 100 million years or so, but notwithstanding previous aborted attempts, the Ice Age did not really get going until just 2.5 million years ago. Clearly, other things needed to happen first, one of which may have been the closure, between about 4.5 and 2 million years ago, of the

Isthmus of Panama. This does not seem much of a big deal, but it was a landmark event that may have provided a critical piece of the jigsaw whose completion would trigger the launch of full ice age conditions. The formation of the isthmus cut communication between the waters of the North Atlantic and the Pacific and resulted in a major reconfiguration of ocean currents. The corollary of this seems to have been a sudden and dramatic change in the climate of the North Atlantic, such that the formation of ice sheets across North America and Europe were promoted, if for somewhat counter-intuitive reasons. Once the isthmus had formed, the warm waters of the tropical Atlantic were no longer able to leak into the Pacific and instead were forced northwards thereby strengthening the Gulf Stream and associated currents that transported warm waters towards the Arctic. Such a change might have been expected to reduce the chances of ice sheet growth at high latitudes, but ultimately the opposite seems to have been the case. As well as moving more heat northwards, the invigorated Gulf Stream also carried more moisture to high latitudes. This, in turn—once all the other pieces of the ice-world jigsaw were in place—would have stimulated snowfall and promoted the growth and expansion of the northern hemisphere ice sheets.

An additional global factor that is key to the establishment of ice-house conditions is the level of atmospheric carbon dioxide, which needs to be relatively low so as to limit greenhouse warming. During the Cenozoic, the concentration of carbon dioxide in the atmosphere followed a broadly downward trend, punctuated by the odd blip, from more than 1000 ppm at the height of the Eocene warmth to below 180 ppm during the frigid glacial episodes of the past million years or so. Why carbon dioxide levels should have fallen as they did is not yet fully understood. One idea, expounded as long ago as 1988 by US scientists Maureen Raymo, Flip Froelich and the same

Bill Ruddiman who linked the rise of agriculture with the start of anthropogenic warming, blames a burst of mountain building during the latter part of the Cenozoic, involving the Andes, western North America, the Swiss Alps, and the Southern Alps of New Zealand, but most particularly entailing the uplift of the Himalayas. Around 40–50 million years ago, after a swift—geologically speaking—journey northwards, India crashed into the underbelly of Asia, pushing up the Himalaya mountain range and the neighbouring Tibetan Plateau— the largest geographic feature on the surface of the Earth. This created an enormous physical barrier that acted to intensify the Asian monsoon as water-laden air masses from the south encountered the new upland area. The increased rainfall is charged by Raymo and colleagues with increasing enormously the rate of chemical weathering so as to suck carbon dioxide out of the atmosphere in huge volumes. Rainwater combines readily with carbon dioxide in the air to form a weak solution of carbonic acid that is very effective at dissolving rocks and minerals. The soluble products of chemical weathering incorporated the carbon, which was carried down the prodigious river systems that drained the great new upland region and eventually dumped in the oceans. Here, the carbon dioxide was ultimately isolated through being incorporated into the carbonate skeletons of tiny marine creatures or squirrelled away in marine sediments. Raymo and her co-researchers argue that the greatly enhanced opportunity for increased chemical weathering offered by the growth of the Tibetan Plateau and the formation of other upland areas during the later stages of the Cenozoic, all of which would attract high rainfall, resulted in carbon dioxide being progressively extracted or 'scrubbed' from the air leading to the observed downward spiral in concentrations of the gas in the atmosphere, thereby making for a cooler Earth on which it was easier for ice sheets to form and survive.

Ice and hippopotamuses

Starting a little more than 2.7 million years ago, during the late Pliocene, our world finally said goodbye to any semblance of a persistently warm climate, as, over a period of a couple of hundred thousand years, the ice sheets started to build across the landmasses of the northern hemisphere, first in the European Arctic and north-east Siberia, then in Alaska, and finally across north-eastern North America. By 2.5 million years ago, the first glaciation of the last ice age had reached its greatest extent and the world had entered a new period, the Quaternary. For most of its length, the Quaternary is coincident with the Pleistocene Epoch. Just under 12,000 years ago, however, this gives way to the Holocene, the most recent epoch of Earth's immense history, which defines post-glacial times and is characterized by a broadly clement climate.

The final straw that tipped the Earth into full ice age conditions at the end of the Pliocene seems to have originated not from any trick of geography or realignment of ocean currents, nor from a modification of the disposition of the Earth's tectonic plates or a change in the composition of the atmosphere, but to the cyclic variation in our planet's somewhat wobbly passage around the Sun. During its journey through space, the Earth wobbles on a range of timescales and in a variety of different ways. These have, for quite some time, been suspected of being in some way linked to the switching on of the Quaternary Ice Age and to the modulation of waxing and waning glaciations that make it up. A link between the most recent ice age and what is known as astronomical forcing was first proposed in the mid 19th century by French mathematician, Joseph Adhémar. Twenty years later, the self-taught Scottish scientist James Croll developed the idea further, but it was not until Serbian engineer Milutin Milankovitch got to grips with the problem in the 1930s that the precise nature of the connection was established and quantified. What is now known as the Croll–Milankovitch Astronomical Theory of the Ice Ages, argues that the repeated retreats and advances of ice sheets that characterize the Quaternary Period are functions of long-term variations in the geometry of the Earth's orbit and rotation.

With everything else in place, getting the last ice age up and running appears to have been dependent upon the obliquity of the Earth's axis, in other words the tilt of our planet's axis of rotation with respect to the plane of its orbit, being just right. Variations in obliquity constitute just one of the wobbles demonstrated by our planet as it travels around the Sun. The changes in obliquity are not random, however, but form predictable cycles that were recognized by Milankovitch as having a key role in the workings of the last ice age. Over a period of 41,000 years the tilt of the Earth's axis changes from a little over 22° to 24.5° and back again, resulting in small, but significant, changes in the amount of sunlight received at the surface. If the Earth's axis were not tilted we would not experience the seasons. As it is, during the northern hemisphere summer, the North Pole is tilted towards the Sun, allowing more direct solar radiation to reach the surface north of the equator, leading to higher temperatures. During the winter, the North Pole is tilted away from the Sun so that the longer, warm days of summer are replaced by months of cold and reduced daylight. Now it is the southern hemisphere that receives more direct sunlight with the result that our antipodean friends bask in relative warmth while we shiver beneath gloomy skies.

A larger tilt will result in more extreme winters but, perhaps counterintuitively, this is not the best way of triggering the growth of ice sheets. In order for an ice sheet to get going, it is far more important for the summers at high latitudes in the northern hemisphere to be cooler, something that is more likely when the tilt of the Earth's axis is smaller. Under these conditions, snow and ice that has accumulated over the winter months has a better chance of surviving through to the following winter, when more snow and ice will be added. As snow accumulates year on year, so the reflectivity, or albedo, of the surface is increased, resulting in the melting potential of summer sunshine being further reduced, thereby accelerating ice sheet growth. Such a situation seems to have prevailed during the late Pliocene, jump-starting the series of glacial advances and retreats that have continued for more than 2.5 million years since. Interestingly, the current value of tilt is 23.4°, a bit more than half way between its two extremes, and is decreasing. Around about 8000 years from now the tilt value will reach its minimum, which will once again provide ideal conditions for the ice sheets to build themselves anew. As discussed later, however, this projection takes no account of the possible effects of anthropogenic greenhouse gas emissions, which could make for a very different world indeed.

Fifty episodes of glacial advance and retreat are recognized during the course of the Quaternary. To start with, these episodes repeated about every 40,000 years leading to a general consensus that the main control was the cyclic variation in the Earth's axial tilt. The continental ice sheets associated with the early glacial advances were not especially large, and the temperatures during both the glacial episodes and the intervening interglacials, not particularly extreme. As time went on, however, the nature of both glacial and interglacial episodes began to change. Between about 1.5 million and 600,000 years ago, the lengths of the glacial–interglacial cycles began to increase until, after this time, they ranged from 80,000 to 120,000 years and averaged around 100,000 years. The transitions from cold to warm and back changed too. While those of the early cycles were regular and symmetrical, later transitions typically involved long periods of cooling of up to 80,000 years, followed by sudden flips to inter-glacial conditions over periods as short as 4000 years. Looked at over the last million years or so, this behaviour results in a global temperature graph that has a distinctive saw-tooth pattern, with repeated gently dipping downward trends terminated by sudden short upswings. The climate also began to vacillate more wildly over this period, so that while glacial episodes lasted longer and temperatures fell lower and more ice accumulated, the interven-ing interglacials were marked by unusually warm conditions, so warm in fact that in the depths of the last ice age we find condi-tions notably hotter than those encountered in modern times. As for conditions prevailing at the PETM and during the Mid-Miocene Climatic Optimum, these warm intervals may provide some use-ful nuggets of information that could shed a little more light on what our future anthropogenically warmed world might look like. The five interglacials of the past half a million years, including the one we find ourselves in now, were also associated with natural atmospheric carbon dioxide levels higher than those characteris-tic of earlier and cooler interglacials. Of course, in the case of the latest—the Holocene—we have already made sure that concentra-tions of the gas have been artificially forced up to far higher levels.

Alongside the PETM and the MMCO, the interglacial immediately before the current one is studied with great interest by climatolo-gists keen to look for possible clues to our future climate. The Eemian Interglacial takes its name from the valley of the Eem river in the Netherlands, where fossilized plant and animal remains pro-vide testament to a warm and pleasant climate. The Eemian began around 130,000 years ago, coinciding with the termination of the penultimate glaciation, and lasted for about 14,000 years, although conditions in parts of Europe seem to have stayed warm for a few

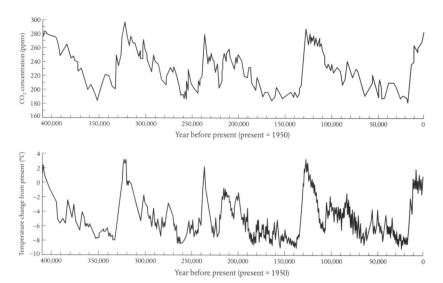

Fig 8. Data from the last 400,000 years show how changes in the Earth's temperature and the atmospheric concentration of carbon dioxide go hand in hand. Over the past 600,000 years or so, glacial-interglacial cycles ranged from 80,000 to 120,000 years in length and averaged around 100,000 years. The transitions from cold to warm and back typically involved long periods of cooling of up to 80,000 years, followed by sudden flips to interglacial conditions over periods as short as 4000 years. This is revealed in global temperature and carbon dioxide concentration graphs that have distinctive saw-tooth patterns, with repeated gently dipping downward trends terminated by sudden short upswings.

thousand years longer. The rapidity with which the climate switched from icehouse to hothouse conditions was particularly astounding, with hippos and other animals from the tropics roaming England and other northern temperature zones just a few thousand years after the landscape was buried beneath ice or frozen solid. At its peak, around 125,000 years ago, there is evidence for considerable recolonisation of the tundra, with forests reaching as far as north Norway and Canada's Baffin Island, both well inside the Arctic Circle. On average, the best guess is that global temperatures were 1–2°C higher than at present, but information from ice cores indicates that Greenland was 5°C warmer, while summer temperatures in parts of Asia may have been up by 4°C compared to today. The Eemian oceans were about 2°–3°C warmer than now and—most worryingly for us—sea levels were significantly higher. Jonathan Overpeck, a palaeoclimatologist at the University of Arizona, and his co-researchers, reckon that global sea levels during the Eemian were a good 4–6m higher than they are today. Overpeck and his team link the rise to partial melting of both the Greenland and West Antarctic ice sheets. Ominously, they also point out that, assuming a 1 per cent annual increase in carbon dioxide or equivalent greenhouse gases, which is only a little higher than we are seeing today—Greenland could end up being far warmer than it was during the Eemian, increasing the likelihood of even more melting and correspondingly higher sea levels. Warming of West Antarctica, and the waters that surround it, may also see its ice sheet making a bigger contribution to future sea level rise than it seems to have done during the Eemian. Equally concerning is the revelation that sea levels during the Eemian may have been rising at a rate as high as 2 cm a year. This is not only close to twice as fast as the average rise that followed the end of the last glacial period, but something like 10 times higher than the rate of 20th century sea level rise

If human emissions continue unabated, the temperatures that characterized the Eemian climate will undoubtedly be achieved by the end of the century, if not considerably sooner. Clearly, then, the interglacial qualifies as a close analogue for an anthropogenically warmed Earth of the near future. For this reason, scientific interest in the Eemian has been sufficient to attract funding and support for a 14-nation research project, led by Danish and US scientists, aimed at drilling down through the frigid wasteland that is north-west Greenland in order to reach and retrieve ice that was formed at the time of the interglacial. While previous drilling projects in Green-land have reached and passed through Eemian ice, the cores extracted have been of poor quality, either due to melting or because of having been disturbed or disrupted by ice flow close to the underlying bed-rock. The location for the North Greenland Eemian Ice Drilling Project (NEEM), has been chosen to provide the best chance of Eemian ice being extracted undamaged. Three years after the drill-ing campaign started, the research team celebrated encountering bedrock more than 2.5 km beneath the surface in late July 2010, pass-ing through and retrieving Eemian ice on the way down. It is too early yet to say whether this ice, which has not seen daylight for more than 100 millennia, will yield up any surprises or surrender

Fig 9. In 2010, the North Greenland Eemian Ice Drilling Project (NEEM) encoun-tered bedrock more than 2.5 km beneath the surface of the Greenland Ice Sheet; passing through and retrieving Eemian ice on the way down. It is hoped that this ice will supply new information about the warm episode immediately prior to our own, so helping us to understand better the climate future that awaits us.

any information that may help us to understand any better the climate future that awaits us. It would be odd indeed, however, if the warm episode immediately prior to our own did not have something left to teach us.

It is doubtful whether most of us give much thought to the last ice age. When and if we do, our imaginings will almost certainly be of a frigid and inhospitable episode in our planet's distant past when mammoths and sabre-toothed tigers battled it out with straggly bearded hominids with flint-headed spears—images in our mind's eye fed by museum dioramas and, increasingly, by what we see on television and the cinema. We are unlikely, however, to contemplate perhaps the two most important points about the ice age. First, that the retreat of the ice sheets happened so recently as to be indistinguishable from the present day on any reasonably scaled representation of the history of our planet. Second, that we are still in it! It is just 20,000 years or so since the Last Glacial Maximum (LGM), when much of Europe and North America lay imprisoned beneath ice more than 3 km thick, global average temperatures were 6° lower than they are today and the world's ocean basins were drained of up to 130 m of water. Perhaps more astonishingly, barely 400 human generations later, temperatures and sea levels had rocketed and the continental ice sheets had vanished, transforming our world into one little different—at least in gross characteristics—from the one we live in today. But we shouldn't be fooled. This does not mean that the ice age is over and the glaciers gone for ever. The Holocene Interglacial, which is defined officially as starting 11,700 years ago, is just the latest of many. Not only are we still firmly ensconced within the ice age, but we might expect the ice sheets to start building again in about 8000 years or even sooner. This, of course, assumes that the hugely inflated levels of carbon dioxide in the atmosphere arising from human activities do not delay the return of the ice. According to Bill Ruddiman, the greenhouse gas emissions produced by early humans as they swapped their spears for seeds may already have staved off an early return of ice to northern Canada. Looking ahead, it has been suggested that so great are the quantities of carbon dioxide we are pumping into the atmosphere that we may skip anything from the next glaciation to the next five. This would mean that it could be another half a million years or more before the great ice sheets are able to form again. In which case, the current Holocene interglacial may well be the longest ever, stretching into an extended period of hothouse conditions to rival or even far outlast the PETM.

Because of the way we are reshaping our planet and our climate, there is a growing argument that our modern world should not be part of the Holocene at all. Clamour is increasing for a new epoch to be established, marking the period over which humankind and its activities have had a significant global impact. The idea of a so-called Anthropocene Epoch was first suggested a decade ago by Dutch atmospheric chemist and Nobel Laureate, Paul Crutzen, but it has yet to become part of the established geological timescale. A clear stumbling block involves defining the starting point of the Anthropocene. The beginning of the industrial revolution in the late 18th century would be an obvious candidate, but this does not take account of the fact that human activities may have begun to alter the composition of our planet's atmosphere and its climate as far back as 8000 years ago or even earlier. This would make the Anthropocene almost coincident with the Holocene and undoubtedly lead to confusion and complication. It might be easier just to attach a rider to the definition of the Holocene, to the effect that this is also the epoch during which human activities resulted in an increasing influence on climate, environment and ecosystems, mostly—it has to be said—to their detriment.

Welcome to the Holocene

Whatever we call our interglacial, and whatever its future, we can at least look back at its first 11,700 years with some knowledge and certainty. Many outstanding questions remain about this post-glacial epoch, but constituting as it does the most recent episode of the 4.6 billion year long geological timescale, we know far more about it than any earlier time. While certainly an oasis of calm in comparison to the frenzied environmental changes of the previous 10,000 years, it would be wrong to think of the Holocene Epoch as unexciting. The climate was broadly stable and as such provided the perfect setting for the advance of human culture and, in particular, the switch from hunting and gathering to farming, and the coming together of small groups of individuals into larger communities. The Holocene climate was still not one to be taken for granted, however, and there were certainly stings in the tail of the preceding Pleistocene Epoch that harked back to a frostier world. In fact, the Holocene was itself heralded by one such sting, which marked the final freezing fling of the Pleistocene. This took the form of a severe cold snap, rather quaintly named the Younger Dryas after a pretty white alpine flower whose pollen is found widely during this chilly episode.

A little under 13,000 years ago, the icy wastelands were largely a thing of the past, sea levels were heading relentlessly upwards towards today's levels and the world's climate was warmer and wetter than at any time since the Eemian. Then, all hell broke loose. Suddenly the climate flipped so that temperatures plunged to depths almost as low as during the height of the last glaciation. In the UK, the average temperature fell by about 5°C, while in central Greenland it dropped by 15°C. For 1300 years, the world shivered, or at least much of the northern hemisphere did. Evidence for this short episode of bitter cold

66

appears to be absent in Antarctica and much of Oceania and, somewhat problematically, looks to have started earlier across South
America. The cooling seems to have been less intense in western
North America than in Europe and the North Atlantic region, which
should not be a surprise, bearing in mind the likely trigger for the
Younger Dryas. While the cooling has been linked to a putative
impact event, the consensus holds that the freeze had a more down-
to-earth origin. By Younger Dryas times, the great continental ice
sheets had been melting for several thousand years, producing colossal volumes of icy freshwater, much of which accumulated in gigantic pro-glacial lakes along the front of the receding ice sheets. The
grand-daddy of them all was Lake Agassiz, a vast body of water that
spread across much of what are now the Canadian provinces of Mani-
toba, Saskatchewan and western Ontario, along with parts of the US
states of Minnesota and North Dakota. At its greatest extent, Agassiz
merged with neighbouring Lake Ojibway forming a prodigious body
of water with an estimated volume of more than 160,000 cubic kilometres and covering an area of more than 840,000 square kilometres,
making it almost three times larger than the Caspian Sea, currently
the biggest inland body of water on the planet. Around 12,800 years
ago, part of the ice and rock debris that enclosed the lake gave way,
allowing more than 9000 cubic kilometres of lakewater to drain catastrophically northwards, into the Arctic Ocean. The effect of tril-
lions of tonnes of cold freshwater cascading into the North Atlantic
seems to have been immediate and cataclysmic, and the prevailing
theory holds that the deluge dramatically curtailed the Gulf Stream,
or even stopped it in its tracks. As a consequence of shutting off this
source of tropical heat, near-Arctic conditions soon returned to
Europe and much of North America, with the cooling gradually moving outwards to affect other parts of the planet. It is worth mention-
ing here that a recent paper offers a modified theory, charging

increased rainfall with the freshening of the North Atlantic that reined in the Gulf Stream.

Whatever its cause, the really extraordinary thing about the Younger Dryas is how quickly it started. Originally, the transition from clement warmth to bitter cold was thought to have occurred in a matter of centuries, impressively short on a geological timescale. More recently, it has been widely held that the flip occurred in just a few decades—even more remarkable. In 2009, however, the results of a rather unglamorous-sounding study centred upon a core of mud from the bed of an ancient Irish lake came up with an even more astonishing timescale. Canadian biogeochemist, Bill Patterson and his team of researchers chose Lake Monreagh in County Clare, in particular, because of the excellent and undisturbed preservation of the mud layers in the lake sediment that extended back to latest Pleistocene times. Individual layers each marked just a few months worth of deposition, allowing the scientists to work out very precisely the timing and speed of any environmental changes recorded in the mud sequence. The team measured carbon and oxygen isotopes to determine, respectively, the level of lake productivity and the temperature of the lake water. What they found left them shocked. At the start of the Younger Dryas, the lake suffered a serious drop in both productivity and temperature—not over a period of decades but in merely 1–2 years, perhaps even over just a few months. This is almost on a par with the speeded-up deep-freeze that brought the world to its knees in the film *The Day After Tomorrow*, and it does smell more of science fiction than science fact. As Bill Patterson himself commented to the media following announcement of the findings a year ago, 'it would be like taking Ireland today and moving it up to above the Arctic Circle'. This is not science fiction, however, but simply the latest piece in the jigsaw of hard scientific evidence that is building a disconcerting picture of just how rapidly the climate can switch, particularly when—as at present—it is severely stressed.

The Younger Dryas was a rather exceptional forerunner of the Holocene, and nothing this climatically dramatic has occurred since. Nevertheless, the broadly benign conditions of the past 11,700 years have been periodically punctuated by further abrupt and short-lived shocks caused by climate flips, albeit on an altogether less impressive scale than that which immediately preceded Holocene times. Lake Agassiz, for example, still presented a threat, and following the Younger Dryas flood it was not long before the lake was brim-full once again. A second flood during the earliest Holocene is held responsible for a cool period lasting a few centuries, known as the Preboreal Oscillation, while another huge deluge from the once again rejuvenated Agassiz is widely accepted to be the cause of another Holocene cold snap known, rather unimaginatively, as the 8.2 ka event (ka being technical shorthand for 'kilo-years' or thousands of years). Like the Younger Dryas, the effects of the 8.2 ka event look to have been felt most strongly around the North Atlantic, although the cooling is recognized across the globe. It seems as if the cold spell was a reflection of the death throes of the combined Lake Agassiz-Ojibway, whose final cataclysmic drain-ing at this time decanted a mind-boggling 160,000 cubic kilometres of freshwater into the North Atlantic, raising global sea levels by 2–4 m virtually overnight. Once again, this seems to have interfered with the system of North Atlantic currents, triggering a 400-year cooling that spread out from the region across the planet. Because the Earth was by this time significantly warmer, the impact of the cooling was not as severe as during the Younger Dryas. Nevertheless, it brought about significant temporary changes to our world's climate, including temperature falls as high as 5°C in some regions, and may also have influenced the course of human history. A resulting centuries-long drought in parts of Africa and Asia, for example, has been held responsible by some for encouraging the development of irrigation in ancient Mesopotamia and with bringing communities together in larger concentrations in order to better cope with resulting food shortages. Certainly, around this time, there seems to have been a sudden change in the demographics of the region, with families and small groups congregating to form the world's first towns and cities.

Although the largest natural climatic blip in the Holocene, the 8.2 ka event is far from being the only one, and is itself considered to be part of a series of brief cooling episodes known as Bond Events—after the late US marine geologist, Gerard Bond—that happen every 1500 years or so and which involve climate fluctuations that originate in the North Atlantic region. A notable climate disruption is not realized at every predicted point in the pattern, but since the 8.2 ka event, they have been recorded at 5900, 4200, 2800 and 1400 years ago, with the latest linked to the so-called Little Ice Age that lasted, roughly, from the 17th to 19th centuries. Bond Events are also recognized earlier in the Holocene, and North Atlantic climate fluctuations that follow a similar pattern are also identified during the preceding Pleistocene, where they are known as Dansgaard–Oeschger Cycles. The cooling that defines the Bond Events seems to be related to changes in North Atlantic currents, clearly augmented at the time of the 8.2 ka event by the catastrophic draining of Lake Agassiz-Ojibway, but the underlying cause of these changes remains to be established. One serious possibility, proposed a few years before his death by Bond himself, links the cooling events to episodic reductions in the Sun's activity. Certainly, the Little Ice Age is known to coincide with a period of reduced solar output known as the Maunder Minimum. More recently, there has also been speculation that the cold UK and western European winters of the past few years are a consequence of the fact that the Sun has recently been less active than for a century or so. Mike Lockwood of the UK's University of Reading, and colleagues, have speculated that reduced solar output can be correlated with the development of persistent winter high-pressure systems across the eastern Atlantic, which act as blocks to milder westerly winds and favour much colder north-easterlies. They have also calculated that there is a close to 10 per cent possibility of a return to Little Ice Age conditions in the next 50 years. The work of Bond, Lockwood and others serves to remind us that the Sun does indeed have a role to play in influencing our planet's climate, and always has done. It is not, however, the driving force of contemporary global climate change, as the skeptics and deniers would have us believe.

In the previous chapter, I made much of the fact that the Holocene provided the climatic and geological stability that permitted the rapid growth and development of human society, and this still holds true. There is also a case to be made, though, for progress to be actively hindered or fostered by changes in the environment. Short-term climate disruption driven by the Bond Event that occurred 4200 years ago, for example, has been implicated in the collapse of a number of ancient societies, including Mesopotamia's Akkadian Empire and Egypt's Old Kingdom; a golden age that saw the construction of the Great Pyramid of Giza and the Sphinx. On the plus side, there is the putative Bond-Event-driven transition from rural to urban living in ancient Mesopotamia. The 5900 Bond Event may also, ultimately, have goaded societal development. This cooling episode is believed to have led to widespread extreme drought, which has been linked to a mass migration of people to river valleys, such as the Nile. As in ancient Mesopotamia, such resulting increases in population density may have helped light the touch paper that eventually stimulated the emergence of more complex communities organized at state level, leading directly to the advent of the great dynasties of Egypt and Sumeria.

Right up until the 19th century, natural variations in the Holocene climate have resulted in shocks and surprises, all of which have had a hand in guiding the development of human society and, ultimately, defining the nature of the world we inhabit and the manner in which we live today. The opportunity provided by the Medieval Warm Period, which lasted from around AD 950 to 1250, for example, was seized upon by the more adventurous Vikings, who established communities in Greenland and Newfoundland. The Greenland communities were destined, sadly, to come to a bad end when the freezing conditions of the following Little Ice Age made life impossible in their isolated northern outpost.

The arrival of the second half of the 20th century has brought with it a truly unprecedented change. Notwithstanding the influences of our distant ancestors on carbon levels in the atmosphere, for the first time, nature has lost its role as the primary determinant of what is happening to our climate. Perhaps bracketing the time that this happened, maybe 1900 or 1950, might prove to be the best way of differentiating the Holocene—during which time nature was still in the ascendency—from the Anthropocene, in which the fate of our future climate rests purely and simply in our hands. The responsibility is mind-boggling. How we live and what we do in the next few decades has the potential to transform not only the world of today and the immediate future, but the world of a hundred millennia and more down the line. Whether we take firm action now to slash emissions or turn our backs on effective action may decide the fates of millions of species and thousands of human generations. We may have wrested control of our future from nature but we can learn a great many lessons from her about the sort of future we have in store, lessons that can help us cope, provided that we take them to heart and act upon them.

Lessons from the past

As I write this, the United Nations Climate Conference in the Mexican resort city of Cancun—known as COP16—has just drawn to a close. Despite managing to avoid a debacle along the lines of that which prevailed in Copenhagen 12 months earlier, and making some progress on peripheral issues, more than 10,000 participants and observers left without achieving any binding global agreement to cut greenhouse gas emissions. Rome is virtually razed to the ground, but Nero is still playing a jaunty little number on his fiddle. Despite some ludicrously optimistic talk in some quarters of keeping the global average temperature rise below 1.5°C compared to pre-industrial times, this is now nothing more than pie in the sky. Given current inaction and the disconnect between what the science says is needed and what politicians and other vested interests feel able to concede, it seems virtually certain that we will not be able to prevent global average temperatures rising by 2°C and very likely much more. Carbon dioxide levels are already well above those of the Eemian interglacial and comparable to those of the Mid-Miocene Climatic Optimum, while the rate at which we are releasing the gas into the atmosphere dwarfs that during the Palaeocene-Eocene Thermal Maximum. Nature may not be in control any longer, but it is sending us warnings from the past, warnings that are currently going unheeded. Looking back over the past 50 million years we have learned that carbon dioxide levels and global average temperatures are intimately linked so that hothouse conditions are invariably associated with high levels of the gas in the atmosphere. We have established that if high carbon dioxide levels and temperatures are maintained then melting ice at the poles will raise the level of the oceans by anything from a few metres to tens of metres. We have determined that the climate system is complex, with feedback mechanisms such as thawing permafrost or decomposing marine gas hydrates having the potential to dramatically enhance warming. We have discovered that the climate does not always change slowly in a linear manner but is prone to sudden flips, sometimes over periods as short as a single year or less, and to sudden transitions triggered at tipping points or discrete thresholds. Perhaps most critically, short-duration climatic disruptions such as the Bond Events demonstrate to us that the Holocene climate that we are messing with is sensitive rather than robust, susceptible as it is to being significantly modified by tiny changes in solar output or in the pattern of ocean currents in the North Atlantic.

All this adds up to the unavoidable conclusion that the scale and speed of human intervention in our planet's climate system is far more likely than not to lead to a return to hothouse conditions later this century; conditions that may well persist for millennia rather than decades. As it has been in the past, full-blown planet-wide rapid climate change is likely—in the future—to be all-pervasive and involve far more than just the atmosphere and the oceans. Sweeping climate change will also act, as it has before, to bend the solid Earth to its bidding; influencing and manipulating once again the geophysical processes that operate at the surface of our planet and in its interior. While not on the gargantuan magnitude that characterized the termination of the last ice age, the loss of ice mass in the polar regions and in high mountain ranges, a rapid rise in global sea levels and a realignment of climate zones and weather patterns have, without doubt, the potential to act together to elicit a vigorous response from the geosphere.

Perhaps the most fascinating and certainly most esoteric of all the connections between climate change and the solid Earth is that encompassing the complex relationships and positive and negative feedbacks that link weather and climate on the one hand with volcanic activity on the other. The means by which large volcanic eruptions can modify the weather and cool the climate is moderately well known and understood; the manifold ways in which weather and climate can persuade volcanoes to erupt, less so. It is a topic that is drawing increasing attention. Read on to find out more.

3

Nice Day
for an Eruption

A word on everyone's lips during April 2010, at least here in the UK and across much of Europe, was Eyjafjallajökull. Mind you, that was normally where it remained, the tongue-twisting Icelandic pronunciation making it near impossible for most English speakers to articulate. What we on the UK government's Scientific Advisory Group in Emergencies (SAGE) quickly came to call 'our favourite volcano' to circumvent the potential dangers of advanced oral gymnastics, had achieved celebrity status, literally overnight, by projecting a huge cloud of ash far beyond Iceland's national boundaries to infiltrate UK and European airspace. Those groups most severely affected had their own—generally unprintable—names for the volcano, most particularly the millions of frustrated travellers marooned or delayed at airports across the world and a couple of apoplectic major airline CEOs, touched with the blarney. For a period of just under a week, the ash cloud brought air travel across the UK and much of Europe to a grinding halt, triggering the largest air-traffic lock-down since the end of World War II. By the time new ad hoc guidelines on safe ash concentrations in the atmosphere had been cobbled together in order to allow passenger aircraft to take to the skies again, the cost of the airspace closure was approaching 3 billion US dollars and more than a few airlines and travel companies were struggling to keep going.

The Eyjafjallajökull eruption supplied a number of salutary warnings, not least that a volcanic event can have a very long reach, threatening even those countries that have no active volcanoes to call their own; a message reinforced by two other aviation-disrupting eruptions in 2011. The blast also provided some measure of how wide-ranging and injurious the impacts of a single, relatively small, eruption can be, even when no death or destruction is involved. Finally, Eyjafjallajökull taught us that volcanic eruptions and the atmosphere are intimately linked, with the latter providing a convenient vehicle by which a volcano's influence can be transmitted far and wide. In fact, the latest ashy visitation from Iceland can be seen to be very small beer when compared to the impacts of earlier eruptions hosted by this land of fire and ice.

The year of the Móðuharðindin

Set astride the Mid-Atlantic Ridge, along which two of the planet's great tectonic plates are slowly inching away from one another, and perched above a hot mantle 'plume' that provides a ready supply of magma to the island's many volcanoes, it is hardly surprising that Iceland is never still. Rumbles and groans, either from lurching

crustal faults or from molten rock forcing its way towards the surface, are ubiquitous and accepted as part of everyday life.

The increasingly powerful tremors that shook southern Iceland
during late spring in 1783, however, heralded a volcanic ejaculation
that was far from commonplace, even in a country inured to the violent Earth in all its forms and guises. As the shaking increased in
intensity during May and into June, local people braced themselves
for another eruption of Grimsvötn, a volcano for the most part buried under the vast Vatnajökull Ice Sheet and held tight in its grip,
which erupted most recently in 2011. As the most active of all Iceland's volcanoes, a new outburst from Grimsvötn would have been a
cause neither for surprise nor concern. This eruption, however, would
be very different, not only bringing the country to its knees, but
reaching out across the Atlantic to convey volcanic mayhem as far as
Europe, North America, and beyond. On 8 June 1783, the Grimsvötn
volcano could no longer contain the pressures exerted from below,
and ripped open to allow the rising magma free rein. Not content
with blasting upwards through the ice to form a column of ash, rock,
and debris, as was typical of the volcano's previous activity, the
magma tore open a 27 km long system of fissures from which spewed
forth a huge flood of fast-moving liquid basalt. As the eruption devel
oped, fountains of lava three times higher than the Empire State Building built more than 100 craters along the fissures, feeding flows that
travelled 30 km in just a few days. Typically, draining the supply of
magma that has accumulated within a volcano prior to eruption takes a
few weeks to a few months, with the average duration of an eruption
being around six weeks. But at Grimsvötn, in 1783, the eruption went on
and on. For month after month lava continued to pour from the fissures,
sometimes at incredible rates comparable to the flow of the Amazon
River. At last, although the activity continued into the following year,
the effusion of lava ended in December, by which time close to

14 cubic kilometres had drained from the Earth, inundating a land area of around 600 square kilometres—more than one-third that of greater London.

Long before this, however, both Icelanders and many millions of others further afield had felt the awful and widespread effects of the outburst, not so much as a consequence of the huge outpourings of lava but of the vast quantities of noxious gases lofted into the atmosphere and dispersed across Iceland and the North Atlantic region. During its course, the eruption pumped out an estimated 120 million tonnes of sulphur dioxide, just about three times the annual output of Europe's current industrial base. Accompanying the sulphur gases were even more unpleasant surprises, most notably eight million tonnes of hydrogen fluoride, an extremely reactive and poisonous compound that played havoc with the health and happiness of the island's inhabitants and livestock. For Icelanders, the gases heralded the arrival of the Móðuharðindin, or 'mist hardships'; a time of famine and death that tested to the limit the island population's strength and resilience. The terror and dread of the time is famously and graphically expressed by local Lutheran pastor, Jón Steingrímsson, in his 'fire sermon':

> This past week, and the two prior to it, more poison fell from the sky than words can describe: ash, volcanic hairs, rain full of sulfur and saltpetre, all of it mixed with sand. The snouts, nostrils and feet of livestock grazing or walking on the grass turned bright yellow and raw. All water went tepid and light blue in color and gravel slides turned gray. All the earth's plants burned, withered and turned gray, one after another, as the fire increased and neared the settlements.

For Iceland and its inhabitants, the legacy of the Lakagígar (literally, Craters of Laki) eruption, or Skaftá Fires, as it is sometimes known, was a brutal one. Fields of grass withered away to nothing, birch trees, shrubs, and mosses succumbed to acid rain, and people complained of a noxious haze across the land that made breathing difficult and caused weakness and a racing heart. Both humans and livestock were poisoned or crippled by skeletal deformities brought about by the fluorine that contaminated drinking water, crops, and animal feed. More than 200,000 sheep, three-quarters of the total stock, were wiped out by fluorosis, along with around 40,000 horses and cattle—half of all those on the island. In the resulting three-year famine, somewhere between one fifth and one quarter of Iceland's 50,000 population was lost to the combined effects of poisoning and starvation.

As ably demonstrated by the order of magnitude smaller Eyjafjallajökull eruption in 2010, clouds of volcanic ash and gas are no respecters of national boundaries. In the manner of the 21st century event, weather conditions prevailing in the summer of 1783 were perfect for carrying volcanic clouds southwards towards the UK and Europe. With the Wright Brothers' first flight more than a century away, there was no airline industry to worry about, and any ash in the cloud presented little or no problem to the populations of the north-east Atlantic region. The huge volumes of sulphur gases held in the cloud, however, were quite a different matter. Across Europe, July 1783 was blisteringly hot, with record temperatures prevailing across the UK and northern and western parts of the continent. While the ongoing Icelandic eruption is unlikely to have been the cause of the baking conditions, the anticyclonic situation that brought the heat proved especially effective at capturing the volcanic cloud, steered eastwards towards Europe by the polar jet stream. Once on the scene, the high pressure associated with anticyclonic weather encouraged the cloud to sink towards the surface and spread out across the region, forming a persistent dry sulphurous fog. On 14

June, less than a week after the start of the eruption, the volcanic haze started to descend across France. Less than 10 days later it had formed a blanket across most of Europe and the UK; by the end of the month it had reached Moscow and the Middle East, and by early July it was recorded across the mountains of central Asia. The haze brought a sulphurous smell and a stifling atmosphere, and much more. Shortly after its arrival, crops and other vegetation started to wither as a mist of sulphuric acid droplets settled across the countryside, and trees shed their leaves as if autumn had come early. People were plagued by headaches, breathing difficulties and asthma attacks, which soon began to take their toll, particularly on the very young, the old and the infirm, and mortality rates in the UK and across Europe started to climb. The summer death toll in France due to the Lakagígar cloud is likely to have topped by a long way the 16,000 who perished in the 2003 heatwave, while in the UK, close to 20,000 people may have lost their lives to the volcanic smog and the bitter winter weather that followed.

As if the tribulations of a sulphurous summer were not sufficient, the volcanic haze also resulted in plunging temperatures towards the end of the year and into 1784. In both Europe and North America, the winter was one of the most severe in 250 years, with temperatures on both sides of the Atlantic down close to 1.5°C. This may not sound like much, but the fact that a sustained global temperature fall of just 5°C is sufficient to plunge the planet into full ice age, puts the figure into perspective. The impact of the cloud also appears to have been manifest further afield. It has, for example, been held up as the cause of a weakening of the African and Indian monsoons. This, in turn, is charged with dramatically reducing the flow of the River Nile, severely curtailing the annual floods that are essential to growing crops along the course of the river and on the delta, and bringing famine to Cairo. The serious chilling effect due to the

Lakagígar cloud persisted across Europe and North America for three years, and resulted in a fall of up to 0.5°C in northern hemisphere temperatures as a whole. The 1783 eruption was, however, far from unique, and the world had less than half a century to wait before the arrival of an even greater volcanic shock.

Mary Shelley's inspiration

Little more than three decades after the lavas stopped flowing on Iceland, a huge blast on the Indonesian island of Sumbawa brought temperatures tumbling once more across the globe and played havoc with the planet's weather. In 1815, while 12,000 km away in Europe the armies of Bonaparte and Wellington jockeyed for position and advantage, a gigantic accumulation of magma within Tambora volcano continued to hunt for a way to the surface. In fact, rumblings had first become apparent three years earlier, but it was not until early April 1815 that the enormous pressures exerted by the rising magma finally breached a plug of long-petrified lava to initiate arguably the greatest volcanic explosion of the past thousand years. The eruption climaxed a week or so later in a series of titanic detonations that obliterated the area around the volcano. Sir Stamford Raffles, at the time occupying the post of British Lieutenant Governor of Java, painted a particularly colourful and illuminating picture of the eruption. Perhaps best known now for lending his name to one of the most iconic venues of Britain's colonial past—Singapore's Raffles Hotel, in the city that he founded—Raffles reported that the culminating blasts hurled rocks 'some as large as the head' more than 10 km from the volcano and could be heard in Sumatra, more than 1500 km distant. Complete darkness covered Java, a good 500 km from the volcano, while destruction of communities around the volcano itself was absolute. From a total population of around

12,000, just 36 seem to have escaped. Worse was to follow, both for the people of Indonesia and—for the second time in 30-odd years—for the inhabitants of Europe and North America.

The suffocation of crops beneath metres-thick deposits of ash, along with the contamination of water supplies, led rapidly to widespread starvation and disease that together took an estimated 80,000 lives on Sumbawa and neighbouring Lombok—far more than at the height of the eruption. While the misery of the local population was immediate, it took some time for the eruption to make itself known on the other side of the Earth. In volumetric terms, the Tambora eruption was more than twice as big as Lakagígar, blasting out around 30–33 cubic kilometres of magma. It was also far more violently explosive, hurling huge quantities of ash and gas high into the stratosphere and on a course for the northern hemisphere.

Jumping on a relatively recent bandwagon that seeks to blame all historical events of any significance on upheavals in the natural world, the Tambora eruption has been fingered by some as the ultimate cause of unseasonably heavy rains in Europe during June 1815, churning up the ground so as to hinder Bonaparte's progress and provide Wellington with an advantage that he eventually translated into victory at Waterloo on the 18th of the month. It is extremely unlikely that even an eruption of Tambora's size could have modified weather patterns so as to influence, within a couple of months, the weather on the other side of the world. There is no doubting, however, the enormous impact the eruption had on the northern hemisphere climate the following year. Known as the 'Year without a Summer', the unprecedented cold and wet conditions during 1816 resulted from the coming together of three factors. The naturally depressed temperatures of the period, during which an unusually quiet Sun spawned the centuries-long chill known as the Little Ice Age; the cooling effect of Tambora's 100 million tonne sulphuric acid veil, draped across much of the planet; and an additional and earlier cooling and climate perturbation arising from a large but unlocated eruption that occurred in 1809, the evidence for which takes the form of a volcanic sulphur spike in ice cores extracted from the Greenland Ice Sheet.

This conspiracy of circumstances provided an enormous shock to early nineteenth century society in the North Atlantic region, including driving what historian John Post has called 'the last great subsistence crisis in the western world'. The spectacularly colourful volcanic sunsets were a boon for the great landscape painter, Joseph Turner, but there was little else that was good about the weather of 1816. So wet and cold was the summer that Mary Shelley forewent hiking in the mountains around Lord Byron's Villa Diodatti on the shores of Lake Geneva, where she was staying, instead—in keeping with the mood of the times—developing the idea of that greatest of gothic novels, *Frankenstein*. Not to be outdone, Byron too helped set the tone, and his poem *Darkness* perfectly encapsulates the volcanic gloom that descended across Europe that summer:

> *I had a dream, which was not all a dream*
> *The bright sun was extinguish'd and the stars*
> *Did wander darkling in the eternal space*
> *Rayless, and pathless, and the icy Earth*
> *Swung blind and blackening in the moonless air*
> *Morn came and went and came, and brought no day*
> *And men forgot their passions in the dread*
> *Of this desolation; and all hearts*
> *Were chill's into a selfish prayer for light*

Notwithstanding its influences on contemporary culture, the Year without a Summer also brought unseasonably cold temperatures, with killing frosts and snows devastating harvests and slashing crop yields in North America, Europe, and China. The prices of staple foods soared in response. Food riots gripped cities across the European continent and famine was severe in Ireland and Switzerland. Perhaps 200,000 deaths occurred across Europe, maybe half as a consequence of the impact of a typhus epidemic on an Irish population weakened by malnutrition. The bleak summer of 1816 was succeeded by an exceptionally bitter winter, reflecting an overall fall in northern hemisphere temperatures of about 0.6°C, from which it took a decade to recover.

The climate connection

In common with the Lakagígar flood basalt eruption, the Tambora experience reinforces the fact that severe consequences arising from volcanic eruptions can extend far from the source, to impact upon other nations and regions. It also underlines the important point that volcanic eruptions are capable of significantly modifying the climate across wide areas and for years at a time. The relationship between volcanic action and the planet's climate is complex, with knock-on effects capable of tweaking the climate and weather patterns far from the source. Interestingly and somewhat confusingly, volcanoes are a source of two species of gas that have opposing effects on the climate. They release significant quantities of the greenhouse gas, carbon dioxide, which acts to warm the atmosphere. The quantity of the gas released by volcanoes is a good 200 times less than that accumulating in the atmosphere due to human activities, and over historical timescales the total amount of volcano-derived carbon dioxide has stayed pretty much constant. Recently, it has been suggested that a burst of volcanism at the end of the last ice age may have helped warm the climate—about which more later—but this remains very much a minority view. Widely regarded as more important from a climate perspective are the huge quantities of sulphur dioxide violently exhaled during volcanic outbursts. The aerosol mist of sulphuric acid droplets formed by mixing of the gas with water in the atmosphere is especially effective at blocking incoming solar radiation, partly through absorption and partly by reflecting it back into space, thus cooling the surface and lower level of the atmosphere beneath the acid veil.

There are other ways too that explosive volcanic blasts can influence weather and climate, often with deleterious consequences. To mention a few: large eruptions in the tropics can actually lead to warmer winters, while big eruptions nearer the poles are capable of causing a failure of the South Asian monsoon. Eruptions may also reduce rainfall in the often hydrologically challenged Sahel region of North Africa and can either reduce or augment El Niños—the climatic events that involve periodic warming of the eastern tropical Pacific and which are able, in their own right, to power meteorological chaos across the planet. Large volcanic blasts are not much good for the ozone layer either, due to destructive chemical reactions involving volcanic chlorine, which results in the depletion of ozone in the stratosphere that can last for years.

Lakagígar and Tambora provide just two of many examples of volcanic events promoting serious, and inevitably detrimental, changes in climate and weather. In 1883, the destruction of another Indonesian volcano—Krakatoa—reduced northern hemisphere temperatures by 0.6°C, while, more recently, notable cooling was observed following the 1982 eruption of Mexico's El Chichon volcano and the 1991 explosion of Pinatubo in the Philippines. Looking back a couple of thousand years, an unusually lively eruption of Sicily's Mount Etna in AD 44 seems to have spawned a cooling sulphurous cloud, with the Greek biographer and historian, Plutarch, reporting a sun that was pale and without radiance and that furnished so little heat that fruits never ripened. Nearer our time, in 1258, a huge eruption, somewhere on the planet, has been implicated in a UK summer even worse than that of 1816, while in 1452, a massive blast at the Kuwae volcano on the South Pacific island of Vanuatu produced a sulphuric acid veil at least as large as Tambora's that seems to have resulted in a comparably severe impact on the climate. John Grattan, a world authority on the Lakagígar cloud, has speculated that the final death toll arising from its knock-on effects across the world may have been as high as six million.

Even if this is true, the influence of the Icelandic event on human society pales against the postulated consequences of the giant Toba super-eruption that, 74,000 years ago, blasted out some 300 times more ash than Tambora. The vast accompanying veil of sulphuric acid droplets shrouded the planet, slashing the amount of sunlight reaching the surface and heralding a global freeze. Modelling studies suggest that temperatures could have fallen by 10°C or more, easily enough to wipe out the tropical forests and make things very difficult for most life on the planet. So severe was this so-called volcanic winter, which persisted for several years, that it has been held responsible by some volcanologists and anthropologists for sparking a human population crash that is recognized around this time, which reduced our race to just a few thousand individuals so that it teetered for centuries on the edge of extinction.

The idea that volcanoes are capable of dramatically altering the world's climate and weather is clearly indisputable, but is the converse possible? Can changes in the climate, the weather, and other elements of the environment, cause or influence the activity of volcanoes? If this can happen, then volcanic activity and the broad environment within which volcanoes exist can be thought of as forming a feedback system, with environmental changes affecting the behaviour of volcanoes, whose eruptions are in turn able to modify the environment. As far back as the 1970s John Chappell, of the Australian National University, proposed a possible link between changing levels of global volcanism and advancing and retreating ice sheets. Round about the same time, James Kennett and Bob Thunell, then at the University of Rhode Island, showed that global volcanic activity was substantially elevated for the entire 1.8 million years of the Quaternary Period, during which time the ice sheets were constantly swelling or shrinking. It would seem, then, that the best place to start looking for hard evidence of climate variations driving volcanic activity is during those episodes of Earth history that saw the greatest and most rapid changes to the environment, namely the ice ages.

Our planet has, in the far distant past, experienced the wholesale migration of the ice out of its polar strongholds on a number of occasions, most irresistibly in the late Pre-Cambrian, between 850 and 630 million years ago, when an icy carapace encased most, if not all, of our world, resulting in so-called 'Snowball Earth'. The ice ages we are most familiar with, however, and that we know most about, are those that have held much of the northern hemisphere in their grip over the past two and a half million years or so. The last of these reached its peak around 20,000 years ago, during the Last Glacial Maximum (LGM); a time that marks the maximum extent of the great ice sheets. Since then our planet has undergone a remarkable transformation. As the climate flipped from deep-glacial to interglacial, so rising temperatures led to the rapid melting and retreat of the continental ice sheets; water flooding into the ocean basins as a result drove up long-depressed sea levels by more than 130 m. Hardly surprisingly, this sudden switch from ice world to water world powered enormous changes in the physical environment which, sure enough, elicited a vigorous and widespread response from the world's volcanoes that saw their activity increase from two to six times above normal background levels. If the evidence for volcanoes altering the climate is irrefutable then the molten outburst from the Earth that followed the retreat of the ice makes the case for major environmental changes stoking our planet's subterranean fires equally watertight.

Volcano storm

It seems somehow appropriate that an age of ice should be followed by one of fire; the solid Earth providing a warm welcome for the return of an interglacial climate and its considerably more amenable atmospheric and ocean temperatures. The last ice age was not a good time for volcanoes, with thick permanent covers of ice—particularly at high latitudes and in volcanic mountain ranges such as those of Kamchatka, the Andes, and the High Sierras of California—suppressing eruptions and enforcing extended periods of dormancy. Not only was it far more difficult for magma stored in shallow reservoirs to reach the surface during glacial times, but where the mass of overlying ice was particularly great it seems that less magma was being produced deep within the Earth, thus reducing the supply to the storage chambers beneath and within the volcanoes themselves.

A misconception widely held is that the temperature inside our world is so high that its interior is largely liquid, with just a thin carapace of solid rock keeping us safe from a seething molten maelstrom extending right the way down to the centre of the planet's core, more than 6000 km beneath our feet. True, the temperature here is a warmish 6000°C—about the same as the surface of our Sun—but the enormous pressure, a staggering 3.5 million times that exerted by the atmosphere at sea level, prevents melting and keeps the Earth mainly solid. As the liquid form of most materials (water being a notable exception) occupies a greater volume than their solid equivalents, high pressures act to suppress melting, even where the temperature is sufficiently high for this to be expected. Within our planet, both temperature and pressure increase with depth, but at different rates, which also vary from place to place. Only in a single thin layer within the Earth's mantle—that part of our planet, between the crust and the core, that makes up 85 per cent of its volume—are pressures and temperatures such that conditions are just right for rock to melt.

This 'Goldilocks' zone, better known in geological circles as the asthenosphere, is actually quite close to the surface—between about 100 and 250 km deep. The asthenosphere is a critical element in the make-up of our planet, forming the partly molten layer on which the Earth's tectonic plates skate, and providing the nursery for most magma. Above the asthenosphere—in a zone of cooler brittle crust and uppermost mantle comprising the lithosphere, temperatures are broadly too low to allow rocks to melt, while below the astheno-sphere pressures are too great. The asthenosphere itself is probably about 10 per cent molten, with the balance so close to melting that either small rises in temperature or falls in pressure will initiate the process. One very effective way of reducing the pressure in the asthenosphere is through the removal of mass at the surface, as may be accomplished, for example, by the rapid melting of the vast ice sheet that covered Iceland at the peak of the last glaciation. This had the consequence of allowing the freed lithosphere to bounce back—a process known as isostatic rebound—raising the level of the surface by several hundred metres, and reducing the pressure on the under-lying asthenosphere so as to promote a melting jamboree.

As the first green shoots of the current interglacial burst forth, so too, at volcanoes held in check by the ice sheets for tens of thousands of years, did the earliest ejaculations of magma. Formerly-icebound volcanoes had a great deal of catching up to do, and nowhere more

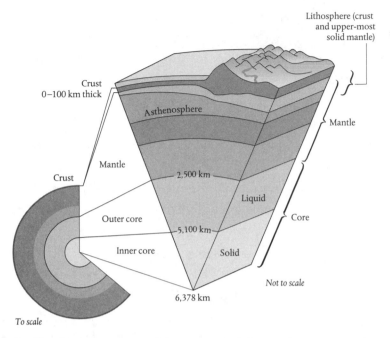

Lithosphere (crust
and upper-most
solid mantle)

Crust
0–100 km thick

Asthenosphere

Mantle

Mantle

Crust

2,500 km

Liquid

Outer core

Core

5,100 km

Inner core

Solid

Not to scale

6,378 km

To scale

Fig 10. The Earth is made up of three concentric layers: the crust, mantle, and core (outer and inner). Within most of the Earth, temperatures are too low or pressures too high to allow rock to melt. Consequently, magma is formed in a 'Goldilocks Zone' known as the asthenosphere, between about 100 and 250 km beneath the surface, upon which the brittle tectonic plates that make up the lithosphere 'float'.

so than in Iceland, home to nine major volcanoes, including the aforementioned Grimsvötn and Eyjafjallajökull, as well as the better known Hekla and Krafla. The key role that volcanoes have played in the formation of Iceland is attested to by the fact that the country's rugged and unique landscape is built entirely from solidified magma—either erupted onto the surface in the form of lava or ash, or emplaced deep beneath the surface to fill the gaps left as the North American tectonic plate, to the west, and the Eurasian plate to the east creep apart about as quickly as fingernails grow.

Twenty thousand years ago, Iceland was entirely covered by a layer of ice that averaged close to a kilometre in thickness. Around 15–16,000 years ago, planetary warming triggered rapid melting of the glaciers, reducing the load acting on the volcanoes beneath and on the underlying asthenosphere. By 12,000 years ago unloading was sufficiently advanced to trigger a spectacular response. Over a period of 1500 years or so, the volcanic eruption rate jumped by between 30 and 50 times, before falling back to today's level. This volcanic rejuvenation was in part a reflection of the release of magma held ready and waiting, within and beneath the volcanoes themselves, but mainly testament to a huge increase in the supply of fresh magma from deeper within the Earth. Such was the load reduction due to the rapid loss of ice mass, that the depressed lithosphere quickly bounced back by as much as half a kilometre, dramatically reducing pressures in the asthenosphere and triggering a 30-fold jump in magma production.

The triggering of volcanic eruptions due to unloading can still be seen in Iceland today, at the now familiar Grimsvötn volcano. Lodged firmly beneath the Vatnajökull Ice Cap, the last remaining major remnant of the ice cover that completely shrouded the island at the height of the last ice age, the volcano is in a somewhat unusual situation. This is made more so by the fact that it does not conform to the simple, cone-shaped structure that fits most people's idea of a volcano. Instead, Grimsvötn, is truncated by a large crater, or caldera, 8 km across, which is lake-filled. Typically, geothermal heat melts the surrounding ice over time so that the lake level progressively rises. Eventually, the pressure exerted by the water bursts the seal with the overlying ice, generating a glacial flood known in Iceland as a jökulhlaup. On occasion, if the volcano is fully charged and ready to blow, then the reduction in the load exerted on the magma reservoir beneath the caldera is reduced to such a degree that an eruption is triggered. This is far from being a rare event, and happened in both 2004 and 2011.

Melting of the great ice sheets at the end of the last glaciation also appears to have influenced the activity of volcanoes located nearby rather than buried beneath them. In western Europe, for example, the volcanoes of the Eifel Mountains in Germany and France's Massif Central seem to have been rejuvenated between about 17,000 and 5000 years ago. The evidence for this is a burst of more than 50 eruptions occurring at a time when the nearby ice sheets that covered Scandinavia, the UK, and the Alps, were undergoing rapid retreat. According to UK geologist, David Nowell, and colleagues, the cause is likely to have been the regional uplift of the crust that accompanied the removal of the great mass of ice. As in Iceland, the effect would have been to reduce pressure in the asthenosphere beneath, resulting in the production of fresh magma and its supply to the volcanoes above. Evidence for the melting of more localized ice fields promoting bursts of volcanic activity also exists for the volcanoes of Russia's Kamchatka Peninsula, the Chilean Andes, and the western United States. The volcanoes of eastern California, for example, reveal an obvious, inverse relationship between fire and ice during the course of almost a million years of glacial advances and retreats in the Sierra Nevada mountains. The likely explanation is that at times of thick ice cover, magma was inhibited from opening and travelling along fractures—known in the trade as dykes—to feed eruptions at the surface. Once ice-free, during interglacial periods, in contrast, the formation of dykes was less hindered so that the eruption of magma was more easily facilitated.

While the volcanic response to the higher temperatures and shrinking ice fields that marked the end of the last ice age was most obviously expressed in Iceland, along with other extensively glaciated volcanic areas such as western Europe, the western United States, easternmost Russia, and the high Andes, it has been shown relatively recently to have been far more widespread. Compiling an accurate catalogue of the world's volcanic activity going all the way back to the last ice age and beyond, and showing how this varies over time so as to highlight peaks of activity, is not an easy task. Detailed accounts of eruptions only cover the past few hundred years, before which reliance has to be placed upon documentary evidence that gets less reliable the older it gets. Further back in time, dendrochronology—a method of dating based upon analysis of tree-ring patterns—can be helpful, particularly in pinpointing large eruptions, which trigger climate cooling and result in poor growth and the formation of narrow rings. Too often, however, such episodes of poor growth could have alternative causes. For specific eruptions at individual volcanoes, dating techniques such as carbon-14 can provide accurate and precise dates, but such methods have not been used extensively enough to build up a picture of how global volcanic activity has changed over time. Fortuitously, the solution to the problem lies deep within the remnants of those great ice sheets that were not lost during the transition to a warmer climate—most particularly the Greenland Ice Sheet.

Four-fifths of Greenland is buried beneath the second greatest mass of ice on the planet, something close to three million cubic kilo metres in volume and up to three kilometres thick in places. As new layers of ice accumulate at the surface of the ice sheet every year so the added weight presses down and compresses those layers deposited previously. This process has been going on for tens of millennia, so that the entire ice sheet is today built from the accrued coatings of more than 100,000 years' worth of amassed ice. This frozen archive provides a unique treasure-trove of information about how the climate changed over the past hundred millennia—as far back as the height of the last ice age—and much further. The water making up the ice, and tiny bubbles of the prevailing atmosphere trapped within, yield invaluable data on all sorts of things, including temperature, the composition of the atmosphere, sea level, and the Sun's activity. The ice also records events that have happened around the world that have dumped large volumes of material into the atmosphere, such as dust storms, wildfires, and volcanic eruptions. The last are expressed in the form of layers of fine volcanic dust or, more commonly, as films of sulphate, left as calling cards and formed by the settling out, across the surface of the ice sheet, of the sulphuric acid aerosol veils created during eruptions such as Lakagígar, Tambora, Krakatoa, and Pinatubo. Translating these sulphate signals into a record of volcanic activity over time, however, has proved to be far from easy. First, the ice had to be drilled to extract cores that could be analysed. Not unexpectedly, given the harsh conditions, this took considerable time to accomplish. Preliminary drilling began in the 1970s and 1980s, but it took until 1993 for both the US-led GISP-2 (Greenland Ice Sheet Project-2) and the European GRIP (Greenland Ice Core Project) programmes to drill to the base of the ice sheet, allowing cores to be extracted that were representative of its full 3 km thickness. Three years later US climatologist, Greg Zielinski, and colleagues, published the first record of volcanism preserved throughout the thickness of the Greenland Ice Sheet as sampled by the GISP-2 core, providing a log of explosive volcanic activity going back an extraordinary 110,000 years. In this seminal paper, Zielinski reports two periods of enhanced volcanic activity, a younger one between 17,000 and 6000 years ago, during post-glacial times, and an older one between 35,000 and 22,000 years ago, when the Earth was cooling and ice sheets growing. Zielinski and his co-workers speculated that these clusters of eruptions might reflect a global volcanic response to the influence of the changing climate at the time, and proposed, as a likely driver, changes in crustal stresses due to ice-mass loss during deglaciation and to ice loading as glaciated areas expanded in the build-up to the Last Glacial Maximum.

Fire and water

The large and rapid climate transitions involved in glaciation–deglaciation cycles involve more, however, than just loading and unloading by ice. There are many other accompanying environmental changes, particularly as the world warms heading into an interglacial, and there are a number of ways that volcanoes can be triggered to erupt at such times. At high altitudes, ice-capped volcanoes will see their glaciers vanish or at least thin dramatically. Unlike beneath the Icelandic volcanoes, released from their 1 km thick icy prison at the end of the last ice age, resulting reductions in load are unlikely to be sufficient to promote the formation of more magma in the under-lying mantle. The loss of ice cover can, though, encourage more explosive eruptions because magmatic gases are freer to expand, tearing the rising magma apart to form a disrupted magma froth that is expelled violently. Furthermore, the loss of the buttressing effect of the ice can make a volcano more prone to collapse, the failure and removal by sliding of part of its flanks triggering an eruption by reducing the pressure on any magma lurking at shallow depths beneath. The stability of a volcano may also be affected by meltwater from the reducing snow and ice cover, or from increased precipitation due to changing weather patterns, saturating the edifice. Such a situation increases the chances of magma and water coming into contact at shallow depths; always a volatile mix and one capable of triggering an explosive blast, either directly or through the promotion of flank collapse.

Also tied up with the wholesale redistribution of our planet's water at times of large rapid climate transition, and commensurate with major variations in ice mass, are comparably impressive changes in global sea levels. The 130 m rise in sea level that followed the end of the last ice age, for example, would have added enormously to the load-related forces pressing down upon the margins of the continents and island chains, along which most volcanoes are located. Could it be that the distinct bursts of activity faithfully preserved in the Greenland ice reflect a volcanic response to changing sea levels around the world, in addition to a reaction to variations in ice load?

In order to check out the possibility, in the 1990s I led a team of European researchers in examining the timing of volcanic eruptions in the Mediterranean region; this time by looking at the record of volcanic ash layers preserved in drill cores extracted from the sea bed. The results proved to be fascinating. There was no relationship between volcanic activity in the region and absolute sea level, but there was a very clear correlation between the *rate* of sea-level change—in other words how quickly it was going up or down—and the incidence of eruptions. As seen in the Greenland ice, the most marked clustering of eruptions occurred during post-glacial times. Between about 15,000 and 8000 years ago, Mediterranean eruptions large enough to leave ash layers on the sea floor took place, on average, every 350 years, compared to only every 1050 years averaged over the entire course of the past 80,000 years. Other clusters of enhanced volcanic activity were also identified, one, from 35,000 to 38,000 years, similar in age to the older cluster recognized by Zielinski and his team, and yet another linked to rising sea levels between 61,000 and 55,000 years ago.

It is sometimes hard to credit just how rapidly sea levels rose, and how relentlessly huge tracts of land were inundated, following the end of the last ice age, and particularly during the very latest Pleistocene and the early to middle Holocene. The average rise in sea level since the Last Glacial Maximum is about 6 or 7 cm a year—about the width of a small hand—impressive enough when you think that global sea level rose just 20 cm or so during the course of the last century. The post-glacial rise was far from constant, however, and took place in fits and starts that reflected the rate at which glacial melt-water made its way from the retreating ice sheets to the ocean basins. During the period of salient volcanic action between 15,000 and 8000 years ago, the world's oceans were subject to three so-called catastrophic rise events, which saw sea levels climbing as much as several metres a century as stupendous volumes of freshwater flooded into the depleted ocean basins from giant meltwater lakes that had accumulated along the fronts of the diminishing ice fields in North America, Europe, and Asia. These new waters swamped low-lying land masses, cut land bridges and trespassed further and further onto the continents. At the same time, they rose rapidly up the flanks of the 800 or so volcanic islands and coastal volcanoes and encroached upon the balance of the world's 1500 active volcanoes, most of which are located within 250 km of a coastline. The enormous weight of added water loaded the crust in a manner that seems to have favoured the expulsion of stored magma, making a significant contribution to the burst of volcanic activity that characterized the post-glacial world. Two things are interesting about the mechanism involved: first, the amount and rate of sea level rise that is needed to trigger a 'primed' volcano does not seem to have been very great at all; second, we can see the same effect happening today at a small but perfectly formed volcano in deepest Alaska.

The strange antics of Paul's volcano

Pavlof fits just about everyone's idea of the archetypal volcano: cone-shaped, mantled in snow and ice, and towering two and a half thousand metres above the barren flat plains of the Alaska Peninsula. It must have presented a spectacular sight to the Russian explorers who first came across the volcano in the mid eighteenth century. While parcelled off to the USA along with the rest of the state for a pittance

Fig 11. Pavlof is just one of 41 active volcanoes in the US State of Alaska. What makes it different is that some of its eruptions in recent decades appear to have been modulated by stress changes associated with seasonal variations in local sea level.

in 1867, the volcano was—as the name suggests—first christened by a Russian: in 1836 one Captain Lutke designating the cone 'Pavlofskoi Volcan', which, according to the Alaska Volcano Observatory, translates broadly to Paul's or St Paul's Volcano. Pavlof is actually one of twins, and along with its companion, the slightly shorter but equally comely Pavlof Sister, occupies a position in a line of volcanic vents that run parallel to the eastern edge of the Peninsula. Another member of the family is also close by, with the smaller volcano Little Pavlof—a younger brother maybe?—resting on its south-western flank.

Tucked away in the back of beyond, and just one of 41 active volcanoes in the state, Pavlof rarely makes even the local news, and only then if an ash cloud associated with an eruption threatens to disrupt the dense air traffic following the polar flight paths across this remote part of the world. Nevertheless, with more than 40 eruptions over the last couple of centuries or so, Pavlof holds pride of place as

Alaska's most active volcano, and one of the most frequently erupting in the USA. The volcano is scientifically fascinating, but not as a consequence of the frequency of its eruptions, rather because of their timing.

Over certain periods in its recent history this is a volcano that has shown itself to be particularly fussy about just when it erupts, eschewing the warmth of the summer months and confining its activity to the colder weather and longer nights of the autumn and winter. During the 15 years from 1973 to 1998, Pavlof erupted 16 times, with 13 of these eruptions occurring between September and December. Statistically, this is clearly a far from random distribution, but rather a pattern of activity that hints at something—either inherent in the volcano itself or in its local environment—controlling or regulating the eruption of magma. But what? Taking up the challenge of unravelling the mystery, US seismologist, Steve McNutt, and colleague John Beavan, compared the timing of eruptions with a range of different phenomena that they thought might shed some light on why the volcano seemed to prefer to erupt during the colder months. No obvious correlation was observed with earthquakes, either deep or shallow, so it didn't seem as if the timing of eruptions was being controlled by any tectonic effect. Broadening their search, McNutt and Beavan examined a number of environmental factors to see if these might be regulating the volcano's behaviour. Neither changes in atmospheric pressure nor in the tidal cycle could provide an explanation, but the researchers came up trumps with annual variations in sea level. During the autumn and winter months, changes in the region's wind patterns result in the level of the Pacific Ocean creeping upwards along the Alaska Peninsula. After correcting for seasonal variations in atmospheric pressure, this rise in sea level amounts to only 17 cm—about the span of an outstretched hand—small, but sufficient, it seems, to have a controlling influence on the timing of Pavlof's eruptions.

It does take a stretch of the imagination to come up with a reasonable hypothesis for a mechanism by which such a negligible rise in sea level might cause a volcano to erupt, although maybe not a great one. After all, water is heavy. One cubic metre weighs a tonne—and even a small sea level rise involves a great deal more of it piling up around the many thousands of square kilometres that make up the Alaska Peninsula. Nevertheless, the extra pressure exerted on the crust by an additional 17 cm depth of seawater amounts to a little under two kilopascals (kPa), which is just one fiftieth that of standard atmospheric pressure at sea level. This is very small when compared, for example, with the variations in sea-level pressure that occur with everyday changes to the weather, which can be five times as great and more. So why don't these influence the timing of eruptions? And what about the diurnal (daily) tides, which involve changes in local sea level of metres rather than just a few centimetres? Why don't these have an effect? The reason is relatively straightforward and involves the rate at which stress is applied and how the material upon which the stress acts responds. In the case of the Earth's crust, the changes in loading stress that accompany the daily tides and atmospheric pressure changes, perhaps as storms pass overhead, happen too quickly for the crust to deform in response. Where the stress is applied for longer, for example due to sea level that is elevated for a matter of months, there is sufficient time available for the crust to distort— in the case of Pavlof enough to somehow encourage the magma to exit the volcano. A good analogy revealing how a material can respond very differently according to how stress is applied is provided by the popular—and very messy—children's plaything known as *Silly Putty*. Give this silicon polymer a sharp blow and it will break; leave it at rest for long enough, however, and it will subside under gravity to form a puddle.

Even so, it was still a bit of a step for McNutt and Beavan to come up with the idea that Pavlof was succumbing to the extra load caused by the added weight of ocean water whose extra depth amounted to a shallow bath-full. Once the idea took hold, however, establishing a possible mechanism was not too difficult. If the extra load exerted caused a small amount of bending of the crust adjacent to Pavlof, the feeder system supplying magma to the volcano would be compressed very slightly, just as happens at the base of a length of rubber tube held at the ends and gently bent downwards, but sufficiently to initiate a squeezing action able to push waiting magma up and out, as McNutt and Beavan put it, 'like toothpaste from a tube'.

If such a small sea level rise can regulate the eruptions of a single volcano in Alaska, then it is not unreasonable to advocate that, scaled up and extended to the hundreds of volcanoes lining many of the coastlines of our planet, exactly the same mechanism might explain a substantial fraction of the extravagant outburst of volcanism that characterized post-glacial times. This is not the only way that rising sea levels may promote eruptions at a primed volcano, and others are discussed later. A second mechanism worth introducing here, however, may also have been particularly important. Computer modelling undertaken by Andy Pullen of Imperial College London has shown that at the same time that rising sea levels alter the stresses within affected volcanoes so as to promote the expulsion of magma, forces associated with ocean loading also act to make the flanks of volcanoes more unstable and prone to collapse. As Washington State's Mount St Helens volcano displayed so well on 18 May 1980, suddenly removing the flank of a volcano—in the form of a rapidly moving giant landslide—is a very effective way of explosively decompressing any magma lurking within. The result is a titanic volcanic detonation, as the contained gases violently tear the magma apart and blast it upwards and outwards. As post-glacial sea levels climbed remorselessly, this may have provided an additional means of facilitating eruptions at coastal and island volcanoes.

The last straw

Returning to Pavlof, McNutt and Beavan, like all good scientists, are careful to point out that correlation does not necessarily imply cause and effect. Nonetheless, the mechanism proposed for Pavlof is convincing, and also provides a practical explanation for how eruptions at coastal and island volcanoes could be triggered during post-glacial times. The case for a sustained seasonal sea level change regulating the eruptions of Pavlof has to be viewed as a strong one. It is strengthened further by an accumulating mass of evidence in support of very small changes in the external environment taking on the role of the proverbial 'last straw', through supplying the final impetus that makes a primed volcano go bang. Stromboli, for example, whose feisty and frequent eruptions have earned it the title of 'beacon of the Mediterranean', is always pretty full of magma, but it seems that its activity becomes even more spirited when barometric pressure in the vicinity is higher, providing that extra little nudge. On the Caribbean island of Montserrat, where the eruption of the Soufriere Hills volcano is now into its 17th year, it is rainfall that seems to supply the decisive push. Eruptions here are related to a giant dome of quietly extruding lava that periodically roars into life following either collapse of part of its flank or as a result of explosions that blast through it. In July 2001, after several months of sustained dome growth during a period of largely dry weather, an episode of torrential rain triggered the collapse of the dome and the formation of pyroclastic flows—deadly fast-flowing torrents of hot rock, ash, and gas. A suggested mechanism is that during very heavy rainstorms, water is able to penetrate into cracks within the dome rather than simply being vapourized at the surface. Here it will rapidly turn to high-pressure steam capable of providing sufficient stimulus to destabilize an already over-steepened dome flank. A similar situation has been recognized at Mount St Helens, where half a dozen lava dome explosions in the 1980s and early 1990s occurred within days of the passage of storms. Here, heavy rain has once again been fingered as the culprit; the triggering of slope instability or the accelerated growth of cooling fractures both presented as possible ways of violently releasing gas from within the dome to drive the explosions.

At Sicily's Mount Etna, my old stamping ground, a rather more esoteric mechanism has been proposed to account for the fact that the volcano seemingly prefers to erupt in late autumn or spring. Exactly why remains to be firmly established, but one suggestion is that its plumbing system is responding to small stress changes within or below the edifice as a consequence of cyclic variations in the velocity of the Earth as it speeds around the Sun. Provided magma is available at shallow depths, fractionally small reductions in stress, at certain times in the cycle, may allow new fractures to open or existing fractures to widen, thereby providing the waiting magma with the extra impetus it needs to reach the surface.

In fact, astronomical factors have long been held up as possible influencers of volcanic activity, and as far back as 1947, the legendary volcanologist and founder of the Hawaii Volcano Observatory, Thomas Jagger, speculated that there might be a relationship between the activity of Hawaiian volcanoes and the phases of the Moon. Certainly, the timing of eruptions at Kilauea, Hawaii's most dynamic volcano by far, during much of the 19th and 20th centuries, seems to have been controlled by the gravitational forces exerted by the Sun and the

Moon. In the same way that they drive the rising and falling of the oceans, the gravitational fields of these two bodies also push and pull solid rock, creating Earth tides that result in our planet being squeezed and stretched by more than half a metre at the equator. Eruptions at Kilauea coincide very well with the highest Earth tides, which occur every two weeks, the likely reason being that stresses acting within the volcano at these times make easier the movement of magma along zones of weakness within the edifice. Earth tides have also been put forward to explain the patterns of eruptions at Stromboli, Mount Lamington in Papua New Guinea, and at the Caribbean island volcanoes of Soufriere on St Vincent and Mont Pelée on Martinique, including—in the last case—the devastating eruption of 1902 that obliterated the town of St Pierre and erased 29,000 lives. Earth tides may, in fact, constitute an effective and widespread driver of volcanic eruptions. In 1973, Fred Mauk and Malcolm Johnston of the University of Michigan examined almost 700 volcanic eruptions since 1900, and decided that there was a statistically significant link between the timing of the eruptions and the fortnightly solid Earth tide maximum, the relationship being demonstrated particularly well for Japanese volcanoes. The overall pattern does not, however, appear to have been that straightforward, with the Earth tide minimum also proving to be a favoured time of eruption for some volcanoes. Inevitably, determining a cause and effect link between Earth tides and volcanic eruptions is not easy. This is because volcanoes are complex beasts that may not be able to react instantaneously to exerted tidal stresses. These may, for example, reduce pressures sufficiently to allow magma to squeeze more easily along pre-existing fractures towards the surface. By the time the magma is extruded—maybe several days later—the link between tides and eruption will no longer be apparent. Bearing in mind such potential lags and other complications, it is surprising that so many volcanoes have demonstrated a link between Earth tides and the timing of eruptions. Perhaps, however, there is an altogether different explanation for the global pattern of volcanic outbursts.

Nice day for an eruption

In 2004, geologists Ben Mason and David Pyle, together with colleagues at the University of Cambridge, published the extraordinary results of a detailed study into the timing of volcanic eruptions over the past 300 years. By examining the start dates of more than 3000 eruptions recorded in Washington's Smithsonian Institute catalogue of volcanic activity, Mason and his co-researchers determined that there was clear and demonstrable evidence of a worldwide volcano season. Just as there are months of the year best suited to conkers, barbecues, and sledging, so it appears there are times in the calendar more favourable for volcanic eruptions. Between November and April, significantly more eruptions start than during the period from April to October.

Remarkable as the whole idea of a volcano season seems, in light of the steadily accumulating evidence for the sensitivity of volcanic systems to even very small changes in the environment, it should really come as no surprise. Clearly, for a tiny change in its physical circumstances to be translated into an eruption, a volcano must be primed, its feeder system critically poised with a supply of magma ready and waiting to come out. The fact that a significant fraction of the world's active volcanoes respond to miniscule seasonal variations in their environment suggests that many are in this state for much of the time. Unlike at the end of the last ice age, no increased production of new magma is involved during the course of a volcano season, nor is there an increase in the number of eruptions overall; rather volcanoes are being 'forced' to erupt within particular time

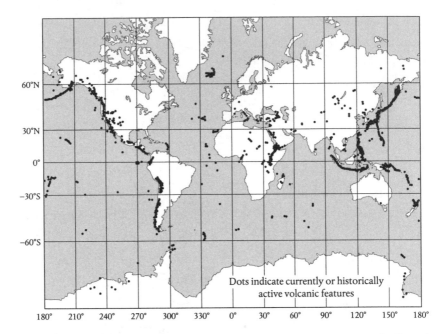

Dots indicate currently or historically active volcanic features

Fig 12. More than 1500 active and potentially active volcanoes are distributed across the planet, their distribution controlled mainly by the disposition of the Earth's tectonic plates. Research on the timing of eruptions at the world's volcanoes, occurring over the past 300 years, suggests that this is not random, with more eruptions than expected taking place between November and April.

frames by external environmental conditions, just as rhubarb is forced to produce edible stems earlier than otherwise, by keeping it in the dark.

Recognising a volcano season is all well and good, but what is the likely cause? Is the observed pattern linked in any way, for example, to variations in the pattern of solid Earth tides, or are large-scale annual changes in sea level or atmospheric pressure somehow part of the equation? In order to check out a link with tidal forces, the Cambridge researchers compared their data set of eruptions with the lunar phases, but could find no statistically significant correlation. So it may be that while certain aforementioned volcanoes, such as the Caribbean's Soufriere (St Vincent) and Mont Pelée, may be prone to the influence of Earth tides, in most cases the associated stress variations seem to be too transitory and too rapidly changing to have any statistically recognisable effect on global volcanic activity in general. It turns out, in fact, that the underlying cause of the seasonality is more complicated, with variations in a number of environmental parameters having roles to play. Ultimately, however, it seems that the seasonal pattern is primarily a consequence of annual deformation at the Earth's surface that accompanies the wholesale redistribution of the planet's water during the operation of the hydrological cycle; the continuous movement of water around the Earth's surface and between the surface and the atmosphere.

Measuring the tiny levels of deformation that may be driving the volcano season is difficult in the extreme and would be impossible without Global Positional System technology. In little more than a decade, GPS has transformed the manner in which we move around or track individuals, and is well on the way to making map reading a lost art. It has, however, been around for far longer than most motorists, walkers, and paranoid parents are aware, and has been used since the early 1990s for accurately and precisely measuring very small movements of the Earth's surface, including the uplift of mountain ranges, the distension of engorged volcanoes prior to eruption and the slow dance of the tectonic plates across the face of our world.

In 2001, UK GPS expert Geoff Blewitt, and his co-authors, published a revolutionary new model in the pages of the journal *Science*, for the wholesale deformation of the planet. Using GPS technology to measure millimetre-scale movements of the Earth's surface, Blewitt's team from the University of Nevada and the UK's Newcastle University was able to recognize a seasonal cycle that involved our world changing shape during the course of a year. The extraordinary result revealed by the study is that, rather like a beating heart, the Earth changes shape systematically and repeatedly, with each 'Earth-beat' taking 12 months. During the course of a single 'beat' the northern hemisphere contracts, reaching a peak in February and March, at the same time as the southern hemisphere expands. This is followed by expansion of the northern hemisphere, peaking in August and September, while the southern hemisphere goes into contraction. The scale of the changes involved is minuscule, and even the most tuned-in Earth Mother would not be able to detect this heartbeat. At the height of the northern hemisphere compression, for example, the North Pole moves downwards by just 3 mm while points near the equator are pulled northwards by just half this amount. The cause of the cycle is fascinating. Throughout the course of a year, a vast wave of ephemeral weight travels across the planet— a conspiracy of changes in snow cover, soil moisture and atmospheric mass. During the northern hemisphere winter, the extra load exerted by lying snow, rainfall-recharged groundwater, and a cooler, denser atmosphere, squashes the crust, the same effect heading south as winter transfers to the other hemisphere. The mass of water shifted from one hemisphere to another during one cycle is colossal, totalling around 10^{16} kg—that is a 10 followed by 15 zeros, or more than one and a half thousand times the weight of all the coal mined in the world in a year.

Taking a leaf from the findings of Blewitt and others, Mason and his co-workers attribute the seasonal responses of the world's volcanoes to the annual deformation of the Earth's surface, driven by the movement of surface water mass during the course of a year. Looked at more closely, however, the different ways in which individual volcanoes or specific volcanic regions behave makes the picture still more complex. Overall, the volcano season is defined mainly by volcanic eruptions occurring around the Pacific 'Ring of Fire' and locally at some individual volcanoes. It is also mainly a reflection of the timing of small explosive events that make up most of the eruption catalogue used. In some parts of the world as many as half the eruptions seem to be controlled by seasonal effects, whereas elsewhere this can be as low as one fifth. Some regions, such as the Mediterranean, show no seasonality at all. Interestingly, and unlike Pavlof, the responses of volcanoes in general to seasonal changes in sea level appear to be ambivalent. In Central America, Russia's Kamchatka Peninsula, and Alaska (notwithstanding Pavlof), seasonal peaks in the eruption rate seem to coincide with times of falling sea level. The island volcanoes of Melanesia, in the south-west Pacific, on the other hand, seem to prefer to erupt when regional sea level is high. The Mason model's bottom line is that while a volcano also responds to other factors and erupts at other times of the year, its eruptions tend to be concentrated—to a greater or lesser degree—at times when the wave of deformation that is constantly traversing our planet in response to annual variations in the hydrological cycle reaches its vicinity.

On chickens and eggs

Why, exactly, is the unveiling of a volcano season important? In many ways, the identification of seasonality in the timing of volcanic eruptions is the icing on the cake—drawing together into a coherent whole previous ideas about how volcanoes respond to external perturbations. As Mason and Co. point out, the discovery that volcanic activity is regulated on seasonal timescales in the modern era provides a contemporary analogue for phenomena previously thought to have only operated many thousands of years ago, during the postglacial period and earlier. Furthermore, the recognition of a seasonal pattern in volcanic activity suggests that previously mentioned responses of individual volcanoes to storms, heavy rainfall or atmospheric pressure changes form just a part of a much bigger-scale relationship between volcanic activity and the global climate. The results of the Mason study also support the idea that episodes of global sea level change in the past were associated with vigorous volcanic responses. But what sense of sea level change is most effective at triggering a fiery burst from within the Earth's interior? Mason's team suggests that falling sea levels might do the job more effectively, while Pavlof and the widespread volcanism that accompanied the melting of the great ice sheets suggest that rising sea levels are most capable of bringing more volcanoes to the boil. Perhaps, as my colleagues and I have suggested, on the basis of our study of Mediterranean volcanoes, it is the rate of change that is important, so that big enough and fast enough changes in sea level—either up or down—will both have the capacity to promote eruptions. The specific mechanisms involved will be different, and there are plenty to choose from. As ably demonstrated by Pavlof, when sea levels are on the up, the resulting crustal loading may change the stresses acting beneath and within an adjacent volcano so as to cajole magma towards the surface. Eruptions may also arise due to collapse of a volcano's flanks, again in response to stress changes within the edifice, to marine erosion, or as a consequence of rising water tables increasing instability and promoting the violent interaction of water and magma at shallow depths. Tensional conditions developed at higher reaches in a volcano, as its lower levels and the crust beneath bend and compress, may allow resident magma to exploit widening and opening fractures to reach the surface. Dropping sea levels, on the other hand, may promote the rise of magma by reducing the weight acting on the flanks of a volcano that is in actual contact with the sea, diminishing the pressure on its storage chamber and allowing magma to push towards the surface more easily. Removing the buttressing effect of the water may also prove effective at triggering eruptions through increasing flank instability and encouraging collapse.

Determining how volcanoes respond to changing sea level is vital in trying to get a better idea of the ways in which volcanic activity and the climate interact. If eruptions can alter the climate, while, at the same time, a changing climate can cause volcanoes to erupt, then we are into a chicken and egg situation fraught with all sorts of potential feedbacks and loops. On the one hand, for example, falling sea levels arising from the growth of ice sheets in a cooling world could trigger a burst of explosive volcanic activity capable of seeding the stratosphere with sulphuric acid droplets, thus propelling the world further and faster along the path to full ice age conditions. On the other hand, the rising sea levels and ice mass loss that characterize a warming world would also be expected to drive a volcanic response, again loading the atmosphere with sulphuric acid aerosols and leading to a cooling effect that could hamper the passage from ice world to water world.

The first situation is an example of a positive feedback, in which the cooling trend promotes the eruptions that then reinforce the trend. The question is, can the strength of any volcanic reaction at times of planetary cooling be sufficient to bring forward the arrival of full ice age conditions? Perhaps it could, if there are sufficient numbers of large explosive sulphur-expelling eruptions, or maybe if there is a single exceptional blast. There has been much speculation about whether or not the great Toba super-eruption, which excavated the world's largest volcanic crater in Sumatra 74,000 years ago, might have accelerated the onset of the last glaciation. As long ago as 1979, Mike Rampino of New York University, and others, speculated that this vast eruption might have been triggered by climate change during a period of global cooling. In the early 1990s, he and Steve Self, now at the UK's Open University, fleshed out the idea that the largest eruption of at least the last 100,000 years may have played a pivotal role at a time of major climatic transition, when expanding ice sheets had already reached about a quarter of the extent of the last glacial maximum. Rampino and Self reflected upon the possibility that the Toba explosion owed its timing to stress changes in the crust brought about by rapid ice sheet growth that resulted in sea levels falling by as much as 40 m in 7000 years.

The statistics of the Toba eruption are hard to comprehend. The eruption was a thousand times greater than that of Mount St Helens in 1980, blasting out close to 3000 cubic kilometres of volcanic debris to heights of up to 40 km or so—around twice the cruising level of the now retired Concord supersonic jetliner—and pumping into the atmosphere up to 5000 million tonnes of sulphur dioxide. As expected, the sulphate signature of the Toba event is easily pinpointed in the Greenland ice cores, which also provide evidence of what happened to the climate in the years immediately following. Global temperatures plunged by 10°C or more, as a consequence of the massive loading of the stratosphere with sulphuric acid aerosols, and remained depressed for several years. This was succeeded by a millennium-length period of extended cooling and glacial advance, blamed by Rampino and Self on the longer-term effects of the eruption. The sudden and impressive temperature fall that immediately followed the eruption is now accepted as convincing evidence of a volcanic winter. Whether or not the extended cold and glacial advances of the next thousand years were a consequence of Toba is another matter. Gareth Jones and colleagues at the UK Met Office demonstrated using a computer climate model that, although snow and ice increased to cover more than one-third of the planet, the initiation of a longer-lasting glacial episode did not seem likely, a result confirmed more recently by US expert on volcanoes and climate, Alan Robock, and his team. Their results reaffirmed the idea that the immediate, rapid fall in global temperatures was a legacy of the eruption, and even suggested that it could have been longer and colder than previously assumed. There was no hint in the simulation, however, that the eruption could have had a sufficiently drawn-out impact on the global climate so as to have promoted a period of glaciation and a further thousand-year cold snap. Nonetheless, Rampino and Self point to other examples over the past two million years, of large, individual eruptions or eruption clusters occurring at times of major climate transition, and flag their possible significance in accelerating cooling or holding back warming, depending on the circumstances.

In the same way that a colossal Toba-like blast may provide a positive feedback at times of planetary cooling, so major volcanic activity occurring during a period of global warming may initiate a negative feedback, such that an increase in sulphuric acid aerosols in the stratosphere may act to hinder warming. This, in turn, could result in the transition from full ice age conditions to the warm climate of an interglacial being more of a struggle than would otherwise be the case. Certainly, the trend in global temperatures during post-glacial times was not consistently upwards, and a number of cold snaps resulted in temperatures falling again, sometimes like a stone. During the 8.2 ka event, introduced in the previous chapter, a deluge of glacial meltwater pouring into the North Atlantic from Lake Agassiz-Ojibway, may have pushed up sea levels by two to four metres, almost instantaneously, with resulting rapid cooling arising from disruption of ocean currents in the North Atlantic. The more severe cold snap known as the Younger Dryas, 4500 years earlier, is also attributed to a great influx of glacial meltwater into the North Atlantic from the same source. It is tempting to speculate, however, that a large eruption or cluster of eruptions during the post-glacial revival of volcanic activity may perhaps have had a role to play, maybe an eruption in

South America, where the Younger Dryas cooling seems to have begun.

Flying foursquare in the face of the idea that volcanoes are instigators of climate cooling, however, is a recent proposal by the Harvard scientists, Peter Huybers and Charles Langmuir, that bursts of volcanic activity at times of major climate change may actually have a warming effect. Today, volcanoes emit between 180 and 440 million tonnes of carbon dioxide every year, a small fraction of the 30 billion tonnes and more resulting from human activities. Notwithstanding this, Huybers and Langmuir argue for a big rise in the amount of carbon dioxide expelled into the atmosphere due to increased volcanic outbursts when the climate is changing dramatically, overwhelming any cooling effect due to sulphuric acid aerosols. If this revolutionary suggestion should prove to be the case, then the relationship between volcanic action and climate change would be turned on its head so that volcanoes helped to accelerate warming at times of deglaciation. While carbon dioxide is produced in not insignificant quantities during volcanic eruptions, the vast balance of current evidence suggests that cooling due to outgassed sulphur dioxide far outweighs any warming effect arising from emitted carbon dioxide. Consequently, the idea remains controversial and at least two studies shove a spoke in the works by demonstrating how volcanic activity is closely correlated with cooling episodes that accompanied the general progressive warming of post-glacial Earth. Unquestionably, far more research and enquiry will be needed to overturn the established tenet, based on a mass of empirical and theoretical evidence, that volcanoes are more effective at cooling our world, rather than heating it up.

Drawing everything together, what can we conclude about volcanoes and their activity in the context of a changing climate? Most crucially, the world's active volcanoes, considered en masse, are revealed to be primed systems that are constantly teetering on the edge of stability and therefore highly sensitive to miniscule changes in their external environment. This allows their activity to be regulated by annual variations in the hydrological cycle today, and made them especially responsive to periods of sudden and dramatic climate change in the past, particularly at times of major climate transition. As anthropogenic climate change drives up global temperatures, and ice mass loss at high latitudes and altitudes is matched by increased mass in the ocean basins; as permafrost melts, weather patterns change, and precipitation characteristics are modified; so major adjustments to the hydrological cycle are certain. It would be astonishing if these changes were not reflected, in turn, in a modified picture of global volcanic activity. Exactly what this new picture will look like and how long it might take to develop are matters best left until the final chapter.

4

Bouncing Back

Mention of Switzerland conjures up either an image of a peaceful and pastoral land of soaring, snow-covered, mountain peaks, deep-blue lakes of icy water, and cuckoo clocks, or one of an industrious and somewhat insular nation of bankers and watchmakers. Whatever picture comes to mind when we consider this small, landlocked country, it is almost certainly not one of seismic catastrophe. Switzerland resides in western Europe far from any tectonic plate margin and away from the more quake-prone Mediterranean region. Despite the occasional tremor as deep, old faults occasionally stir, it must surely be completely immune to any violent shaking of the Earth's crust. This is undoubtedly the widely held view but, as the inhabitants of the city of Basel (or Basle) were to discover to their great cost a little over 650 years ago, it is a perspective that is substantially wide of the mark.

Seismic devastation at the heart of Europe

Today, Basel is a transnational and cosmopolitan urban sprawl of more than three-quarters of a million people straddling a nexus where the French, German, and Swiss borders come together. It is a major business centre and host to the mammoth Swiss pharmaceuticals industry. Basel holds a strategic position on the River Rhine and has been settled at least since Roman times, when it was known as Basilia. By the 14th century it was already an important centre of trade and commerce, guarding one of the few Rhine bridges and hosting the impressive Münster cathedral. Basel was clearly a city that was going places. On 18 October 1356, however, something happened there that sent shockwaves—both real and metaphorical—through western Europe; something that stopped the growth and development of the city in its tracks, at least for a while. Some time between seven and eight in the evening the citizens of Basel felt the ground shake as a small earthquake struck the area. Initial panic quickly faded as the tremor dissipated, but then reappeared as far more severe shaking returned with a vengeance a few hours later. This second earthquake, which may have registered as high as 6.9 on the Moment Magnitude Scale (which has ousted the Richter Scale as the routine measure of the size of large earthquakes)—not that much smaller than the devastating 2010 Haiti earthquake—flattened the city and was felt as far afield as Paris and Prague. According to contemporary reports, 'no church, tower, or house of stone in this town or in the suburb endured'. In addition, countless other buildings, constructed from wood, appear to have succumbed to fires ignited by discarded torches and toppled candles. According to the Chronicle of Basel, the quake killed around 300 people in the city; undoubtedly many others would have been injured as a result of building collapse and the widespread fires, while further fatalities are likely to have occurred outside the city walls. The number still seems rather small, however, considering the apparent scale of the devastation and it may be that the evening foreshock drove people to camp outside their homes or even flee the city altogether. At the very least, they would likely have been sensitized to further tremors and ready to make quick exits from their homes at the first sign of renewed shaking. It is always difficult to reconstruct accurately the intensity and pattern of ground movement associated with an earthquake striking long ago, for which the historical records are typically poor and incomplete, and often simply inaccurate. In this case, however, researchers have been gifted with detailed accounts of the damage sustained by more than 30 castles in Basel and the surrounding area. Analysis of these accounts suggests that the shaking in Basel itself was very severe—reaching intensity 9 (out of a possible 12) on the MSK Seismic Intensity Scale, which is used to estimate the level of shaking caused by an earthquake. This degree of shaking is defined as 'destructive' and today would result in the collapse of poorly built structures and serious damage to well-constructed buildings, cause breakages in underground pipes and lead to ground cracking and widespread landslides. In 1356, it proved perfectly sufficient to obliterate much of the city and neighbouring settlements.

There has been much debate about the precise location of the 1356 quake. The city of Basel occupies a geologically complicated position at a point where stresses stretching the crust along a zone of weakness known as the Rhine Graben meet collisional forces linked to the continuing uplift of the Alps in response to Italy's march northwards. Discussion and deliberation has focused upon whether the 1356 quake resulted from rupturing of a compressional (or thrust) fault to the south of the city, associated with Alpine deformation, or was sourced on a tensional (or extensional) fault in the Rhine Graben to the north—an ancient rift along which western Europe is slowly being

Fig 13. The general perception is that north-west Europe is immune to large, destructive earthquakes. In 1356, however, a quake estimated at magnitude 6.9 obliterated the Swiss city of Basel.

pulled apart. Although not completely resolved, there is now some consensus that the event arose from rupturing of a fault within the graben. While unique in modern(ish) history, it seems that the 1356 quake is by no means unprecedented, with evidence for at least two other comparable events in the past 10,000 years identified by geologists during excavations across the fault. Furthermore, Basel is far from the only place in western Europe that has been shown to be more at risk than previously thought from large, potentially destructive, earthquakes. The Rhine Graben has hosted numerous others, including one in 1992—a magnitude 5.8 event at its northern end, close to the Netherlands town of Roermond, which resulted in damage totalling around €100 million. Fault excavation studies have revealed that three larger quakes shook the region within the past 20,000 years or so, while research elsewhere in and around the northern segment of the graben suggests that the area is crisscrossed by faults that are perfectly capable of generating earthquakes of substantial size. The seismic threat presented by the Rhine Graben may even reach as far as the UK, with individual faults extending westwards beneath the Dover Straits hosting a damaging earthquake in 1382 and again in 1580. The more recent event, which is gauged to have been around magnitude 5.5, resulted in damage across southeast England and is held responsible for two deaths in the capital. Things have been quiet here over the past 500 years or so—some might say ominously so—and there is certainly a reasonable probability that a comparably sized quake will rattle the region in coming decades. As Roger Musson of the British Geological Survey recently pointed out, with the population of London alone now 50 times greater than in the 16th century, a repeat of the 1580 event would be sure to impact upon far more people in south-east England. Loss of life would be greater, while the cost of damage would likely run into many billions of pounds.

What goes down must come up

All the evidence suggests then, that despite its remoteness from active plate margins, the steadfast heartland of Europe is nowhere near as stable as most of us would like to think. Furthermore, there is support for the region being more seismically active in recent millennia than during earlier times. In central Switzerland, for example, a burst of seismic activity seems to have accompanied the transition from the late Pleistocene to the Holocene when the northern hemisphere ice sheets were in full retreat and when the Alpine glaciers were also receding fast. Could it be that the destruction of Basel has its roots, ultimately, in the reduction of ice mass that characterized the switch from ice world to clement Earth thousands of years earlier? Is it possible that the removal of the ice may somehow have permitted the release of stresses that had accumulated on previously imprisoned faults, triggering earthquakes in Europe in a manner similar to the way in which ice unloading launched a burst of volcanic activity in Germany and France following shrinkage of ice cover? There is certainly plenty of evidence to support such a theory, to the extent that it was first proposed more than 80 years ago to explain the pattern of post-glacial earthquakes in northern Canada, Europe, and the UK. The key mechanism underpinning this idea is isostatic rebound, in other words the bouncing back of the planet's lithosphere following its extended detention beneath a thick icy carapace; a vigorous response that fosters the rupturing of long-subdued faults and the consequent generation of some serious earthquakes. To appreciate how and why this happens let me digress briefly in order to provide a short explanation of the concept of isostasy.

In the simplest terms, isostasy is the principle of buoyancy applied to the outer layers of the Earth. Formulated, apocryphally, during his bath-time eureka moment by the Greek mathematician and inventor, Archimedes, the principle when applied to objects floating in water is straightforwardly explained in terms of their displacing their own weight in water. So, a heavier rubber duck, for example, will displace more water than a lighter one of the same dimensions by floating at a lower level. In essence, the same applies in the outer layers of our planet, in which the relatively low density brittle lithospheric plates 'float' on top of the denser plastic, and therefore more easily deformable, asthenosphere that underlies them. In a comparable manner to rubber ducks in a bath, the lithosphere, at any particular point, will float at a level that is consistent with its thickness and density. Massive mountain ranges, for example, will not only soar upwards to tower over the surrounding topography, but will also be underlain by lithosphere that pushes down far into the asthenosphere beneath. Contrastingly, where surface relief is low so the depth to which the lithosphere penetrates into the asthenosphere will be much less. In ideal circumstances, a state of gravitational or 'isostatic' equilibrium exists between the floating lithosphere and the underlying astheno-sphere, but on a planet as dynamic as the Earth this is often not the case. Erosion of mountain belts, for example, may result in a signifi-cant loss of mass, causing them to pop-up in compensation and float at a higher level. Elsewhere, the accumulation of eroded material may add mass to the lithosphere causing it to sink further into the asthe-nosphere below.

Where the lithosphere is loaded by kilometres-thick ice sheets, a state of isostatic disequilibrium exists, such that the colossal weight of the ice forces it downwards to a level that is out of balance with its mass, a situation akin to pushing a floating rubber duck further down into the bathwater. When the ice finally melts, this situation is reme-died by the lithosphere bobbing back up—the aforementioned iso-static rebound—albeit at a far slower rate than the analogous duck. Isostatic rebound following the melting and retreat of the ice sheets is termed, more specifically, post-glacial rebound, to distinguish it from broader isostatic uplift of the lithosphere, which may occur for a variety of reasons. A more accurate term is, in fact, glacial isostatic adjustment, as responses other than simple rebound also occur due to ice-mass loss, for example the loading and sinking of previously exposed continental margins due to post-glacial sea level rise.

At the height of the last glaciation, northern hemisphere ice sheet thicknesses across North America and Europe peaked at an impres-sive 3 km. The colossal weight of these ice masses pressed the under-lying lithosphere further down into the asthenosphere, forcing the hot, plastic material of the uppermost mantle sideways. As rising temperatures led to the rapid loss of ice mass, so this unstable situa-tion started to rectify itself. Initially, the lithosphere started to pop up relatively quickly as it flexed elastically in response to the reduced load above, in a manner similar to a compressed spring regaining its natural length. Later, continued longer-term uplift was controlled by mantle material flowing back, as the lithosphere maintained its upward progress. Due to the high viscosity of the mantle (think treacle, but several orders of magnitude thicker) this was a very slow process, and while the lithosphere confined beneath the ice has since risen by several hundred metres in places, the situation is still not in balance and the process of uplift continues over much of the northern hemisphere's land area, particularly across Canada and Scandinavia. Initially, the rate of uplift was as high as several centimetres a year in areas previously covered by the thickest ice, falling over time to a few centimetres annually, and today to 1 cm or even less. Nevertheless, it could be another 10,000 years or so before the lithosphere in previously glaciated areas has risen to a position that it feels comfortable with.

Physical evidence for rebounding lithosphere during post-glacial times abounds, taking the form, for example, of elevated lake and marine shorelines in previously glaciated areas and patterns of uplift that are not otherwise explicable. In Sweden, close to the heart of the rebound arising from the disappearance of the huge Fennoscandian ice sheet centred on Scandinavia, progressive uplift isolated an arm of the Baltic Sea from the marine environment as recently as the 12th century, transforming it into sweet-water Lake Mälaren on whose shores the nation's capital, Stockholm, now resides. It is only a matter of time before the northern part of the Gulf of Bothnia, which itself makes up the northern arm of the Baltic, experiences the same fate. At Kvarken, Sweden and Finland are only around 80 km apart and the water has a maximum depth of just 25 m. At current rates of uplift—which are a little under 1 cm a year—it would still, however, take more than 25,000 years for the northern part to be cut off. With the uplift rate set to decline over time and with global sea levels rising due to anthropogenic climate change, there is likely to be an even longer wait.

On the other side of the Atlantic, Lake Ontario seems to have been formed in a similar manner. Following melting of the Laurentide ice sheet, the sea flooded into the St Lawrence valley, within which what is now Lake Ontario formed a bay. Rapid uplift soon separated the bay from the ocean, however, forming the freshwater lake we see today. Much continues to happen in the region, with uplift continuing rapidly enough to tilt the lake-bed southwards, transforming river valleys into bays and driving increased rates of erosion that has many shoreline property owners worried. On the other side of the Atlantic, post-glacial rebound continues to cause the UK to pivot across its middle, as previously icebound Scotland and northern England continue to rise, while the south subsides. As a consequence, while parts of Scotland could rise by a further 10 cm or so by 2100, south-east England could have sunk by 5 cm, exacerbating the problem caused by global sea level rise. The occurrence of widespread post-glacial rebound is an established fact but is there clear evidence for a link between the resulting uplift of the lithosphere and increased earthquake activity? Where better to look than in the vicinity of the biggest ice sheet of them all—the Laurentide ice sheet of North America?

The mystery of New Madrid

Like western Europe, the putatively stable heartland of North America has also hosted a seismic surprise or two in recent times, far distant from shaky California and the other earthquake-prone states along the Pacific coast and in Alaska. Most notable were the severe jolts that pierced the tranquil world of the upper Mississippi valley in the early decades of the 19th century. A little over a week before Christmas in 1811, a previously unknown patch of land located between what are now the major cities of St Louis (Missouri) to the north and Memphis (Tennessee) in the south was to achieve some sort of notoriety. On 16 December a major earthquake of somewhere between magnitude 7 and 8, located in north-easternmost Arkansas, shattered the calm of the small hours. Damage to buildings was minimal, mainly because the sparse population required few of them. In the tiny Missouri settlement of New Madrid, located on the Mississippi river a little way north, however, the ground shaking was powerful enough to cause collapse of the river banks and to topple trees. At the spot where Memphis now stands, shaking is estimated to have reached intensity 9, comparable to that which levelled Basel four and a half centuries earlier and certainly sufficient to cause massive damage today. One New Madrid eyewitness—Eliza Bryan—provides a small inkling of the terror experienced by the settlement's citizens, as she recalls in her written account that:

> ... we were visited by a violent shock of an earthquake, accompanied by a very awful noise resembling loud but distant thunder, but more hoarse and vibrating, which was followed in a few minutes by the complete saturation of the atmosphere, with sulphurious [sic] vapor, causing total darkness. The screams of the affrighted inhabitants running to and fro, not knowing where to go, or what to do—the cries of the fowls and beasts of every species—the cracking of trees falling, and the roaring of the Mississippi—the current of which was retrograde for a few minutes, owing as is supposed, to an irruption in its bed—formed a scene truly horrible.

Unfortunately for Eliza and her neighbours, this was just the beginning of one of the most extraordinary seismic series in US history. Powerful aftershocks followed the main shock every several minutes until, around breakfast time, a second major quake—just as big as the first struck. More aftershocks followed but by mid January 1812 the local feeling must have been that they were over the worst—not a bit of it. Another major quake struck on 23 January and yet another—possibly the biggest of all—on 7 February. This final event severely damaged many houses in St Louis and opened ground ruptures that formed instantaneous, albeit temporary, waterfalls on the Mississippi. Because of the efficiency with which the cold, brittle crust of the continental heartland of America transmits seismic waves, the series of earthquakes agitated church bells in Richmond, Virginia, more than 1000 km distant and shook windows in Washington DC, a good 1400 km away, while the quakes were reportedly felt in New York and Boston. Although the February event represented the climax of the sequence, aftershocks strong enough to be felt continued for more than five years.

The origin of the New Madrid earthquake sequence remains enigmatic and continues to be the source of debate and discussion among US seismologists. The fact that the area is about as far from an active plate margin as one can get in the USA suggests that plate tectonics is not directly responsible, so geologists have had to look elsewhere for an underlying cause. Almost inevitably, attention has been brought to bear on the great Laurentide Ice Sheet that buried much of the northern part of the continent at the Last Glacial Maximum; on the influence of its removal on the underlying lithosphere; and on its potential for triggering earthquakes—even many thousands of years after its retreat. One bone of contention with respect to a role for deglaciation in the triggering of the New Madrid quakes of the early 19th century, and other smaller seismic events since, was that the New Madrid Seismic Zone (NMSZ), as it has come to be known, was never actually buried beneath the Late Pleistocene ice. In fact, even at their maximum extent the ice sheets ground to a halt hundreds of kilometres further north. The reach of the ice and its influence on stress and strain in the surrounding rocks, however, is long. When a thick ice sheet forms, the lithosphere immediately below is forced down, as noted earlier, pushing the underlying mantle sideways to make room. Around the margins of the depressed area, the lithosphere flexes upwards in compensation, helped by the outward flow of the mantle, so that the hollow formed beneath the ice is surrounded by a contrasting 'forebulge' that extends well beyond the ice sheet front. As the ice disappears during deglaciation, the lithosphere rebounds while the forebulge gradually subsides in a wave that moves outwards away from the area of the former ice sheet. Ten years ago, this led UK seismologist, Robert Muir-Wood, to venture the suggestion that such a wave of forebulge collapse might have provided the perfect conditions for the release of tectonic strain accumulated during glaciation, resulting in a corresponding cascade of seismic activity that migrated progressively away from the former margins of the ice sheet. As applied to North America, Muir-Wood suggested, this model could explain not only the events in the New Madrid Seismic Zone but also a large and damaging earthquake that struck Charleston, South Carolina, in 1886, and others off the coast of eastern Canada. Patrick Wu of Canada's University of Calgary and Paul Johnston of the Australian National University in Canberra have similar ideas about a wave of seismicity rolling away from the margin of the Laurentide Ice Sheet over time, and have developed a model that seeks to predict the arrival of this pulse of earthquake activity at specific locations in North America. Interestingly, their model accurately predicts the nature and timing of prehistoric earthquakes that occurred in Quebec and Indiana, but has a problem with the New Madrid sequence. While it does predict the timing of the New Madrid quakes, the rate at which the rebound stresses decay with increasing distance

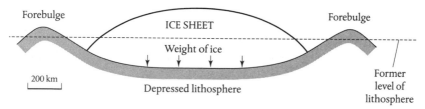

Fig 14. The influence of thick ice sheets on stress and strain in the surrounding rocks is widespread. When an ice sheet forms, the lithosphere immediately below is forced down, pushing the underlying mantle sideways to make room. Around the margins of the depressed area, the lithosphere flexes upwards in compensation, helped by the outward flow of the mantle, so that the hollow formed beneath the ice is surrounded by a so-called 'forebulge' that extends well beyond the ice sheet front. As the ice melts during deglaciation, the lithosphere rebounds while the forebulge gradually subsides in a wave that moves outwards away from the area of the former ice sheet.

from the former ice sheet edge means that they would not be sufficiently large to trigger earthquakes of magnitude 8 in this area. In 2010, however, Susan Hough of the USA Geological Survey scaled down the likely sizes of the New Madrid quakes to a level—around magnitude 7—that can be explained by post-glacial rebound.

Although there seems to be good evidence for the melting of the ice sheets across North America triggering a seismic response, whether the 1811 and 1812 New Madrid quakes were part of this response or not remains an open question. Perhaps the most scientifically reasonable explanation for the New Madrid earthquakes is that both tectonic and deglaciation forces played a role, with the stresses caused by the bounce-back of the lithosphere being just sufficient to nudge faults into life that had already been weakened by the gradual accumulation of stresses associated with the endless passage of the Earth's tectonic plates across the face of the planet. Recent GPS campaigns designed to measure crustal movements so as to estimate how rapidly strain is accumulating today on faults in the NMSZ have suggested that this is occurring at an extremely slow rate, resulting in fault movements amounting to just a few tenths of a millimetre a year. This is more than 100 times less than annual movement due to accumulating strain on California's San Andrea Fault. This information has been used to play down the likelihood of another big quake in the area. It has even led to suggestions that more recent earthquakes in the zone are simply very late aftershocks related to the crust settling down again following the early 19th century events, or—alternatively—to the idea that the NMSZ may actually be 'shutting down' altogether.

The United States Geological Survey (USGS) is sensibly adopting a cautious approach and continues to regard the possibility of another major earthquake striking the area as one that merits concern. As the organisation rightly points out, a decade or so of strain measurements should not be allowed to override a record of persistent seismic activity in the NMSZ that goes back more than 4500 years. The USGS notes that the area continues to be seismically active, with more than 4000 earthquakes detected instrumentally since the mid 1970s and at least one a year large enough to be felt. Furthermore, there is no sign that activity is diminishing over time, which would be expected if continued tremors simply reflected post-1811/12 adjustments in the crust. Given the possible consequences of a future large earthquake in the NMSZ, the USGS is right not to downgrade the perceived level of seismic hazard for the area. The US Federal Emergency Management Agency has warned that a future large quake here could result in the highest economic losses of any natural disaster in US history, outstripping the $80 billion cost arising from the impact of Hurricane Katrina on New Orleans in 2005. Currently, the USGS estimates that the probability of a New Madrid earthquake comparable in size to one of the early 19th century events is 10 per cent in the next 50 years, high enough to ensure that buildings are suitably constructed but not high enough to trigger a mass exodus from the area. To make things more interesting, Robert Muir-Wood has suggested that the wave of post-glacial strain release might keep going, in time moving further south to increase the risk of large earthquakes in parts of Georgia, Arkansas, and northern Louisiana.

Stirred and shaken

Where then do we stand on the relationship between deglaciation and earthquake activity? While it may not yet be possible to pin down exactly the role of deglaciation in triggering of either the 1356 Basel earthquake or the 1811/12 New Madrid seismic sequence, there is strong evidence for a link between ice sheet melting, the rebounding of the lithosphere and increased levels of earthquake activity. Broadly speaking the enormous ice masses that spread across Europe and North America during the last ice age stabilized active faults beneath and suppressed earthquake activity. This effect may be reflected today in the low levels of seismicity that characterize Greenland and Antarctica, although it is also perfectly possible that this is a consequence of a paucity of earthquake faults in the underlying crust.

There is a powerful consensus that the melting of large ice sheets leads to a combination of spectacular uplift and a period of increased earthquake activity as tectonic strain that has accrued on faults beneath and adjacent to the ice sheets is released. Probably the best evidence for this relationship comes from Lapland, the transnational wilderness region that stretches across Arctic Scandinavia and Finland. Here, where the Fennoscandian Ice Sheet was at its thickest, the relationship between faulting and surface features, such as meltwater channels that characterize ice wastage, indicates that major faults moved within just a few years to a few decades of being uncovered by the melting ice—a virtually instantaneous response on a geological time scale. The sizes of the faults, which are up to 150 km long and which may have shifted by as much as 15 m in one go, suggest that they would have sourced some enormous quakes, certainly in excess of magnitude 8. Similar faults have been reported from Canada, although not on the scale of the Lapland structures. This is likely to be a reflection of the fact that rebound stresses following loss of the Fennoscandian Ice Sheet seem to be higher than those associated with melting of the ice that buried much of North America.

Gleaning evidence from beyond the limits of the ice sheets for an indisputable increase in earthquake activity associated with deglaciation is trickier because it is more difficult to fix relationships between the earthquake faults and surface features associated with melting of the ice. Geologists hunt for faults that cut late Pleistocene or Holocene sediments or landforms associated with the so-called 'periglacial' environment that characterized the land areas surrounding the ice sheets, but getting accurate measures of the ages of faulting, and determining whether the cause was tectonic or a product of glacial rebound, has proved to be extremely difficult. Another way of identifying increased earthquake activity following deglaciation is to look within sequences of marine or lake sediments for signs of disturbance that testify to vigorous shaking. Unconsolidated sediment can be churned up by quite moderate earthquakes, as small as magnitude 5, and may thus provide one of the best recorders of earthquake activity in post-glacial times. Indeed, deformed lake sediments in Canada, Sweden, and Scotland, deposited during the last stages of the ice age, or just after, have all been held up as further evidence of a healthy seismic response to wholesale melting of the ice sheets.

A shaky inheritance?

For those concerned with evaluating the current and future risk of earthquakes, particularly in the relatively stable interiors of Europe and North America, the key question is: to what extent, if any, is the post-glacial bounce-back of the lithosphere reflected in the current pattern of fault movement and earthquake activity? In other words, are the heartlands of Europe and North America shakier today than they otherwise would have been as a consequence of what happened at the end of the last ice age? It is far from surprising that it was recent interest in getting a better handle on earthquake hazard and risk in these regions, partly driven by insurers and partly by those responsible for the siting and security of nuclear power plants, which motivated research into the influence of ice-mass fluctuations on earthquake occurrence today, but everyone would like to know what the future holds for these densely populated regions that are ill-prepared to cope with serious earthquakes. As the debate over the ultimate driver of the New Madrid quakes reveals, the answer is not straightforward, with both tectonic and glacial rebound forces likely to be involved.

Robert Muir-Wood is convinced that the interplay between tectonic and rebound-related stresses can explain the patterns of contemporary earthquake activity around the margins of former ice sheets, including those that covered the UK. His elegant idea is that in some places, surrounding a now vanished ice sheet, the forces associated with glacial rebound reinforce the background stresses caused by the movement of tectonic plates, thereby boosting the level of seismic activity. In other places—sometimes known as 'shadow zones'—the two forces work against one another with the result that earthquakes are suppressed. This interplay results in characteristic patterns of earthquake activity, both in the immediate area of the rebounding lithosphere and in the surrounding forebulge, within which two seismically active zones, or 'quadrants', alternate with two quiescent quadrants wherein relatively little is going on seismically speaking. In the region beneath and surrounding the site of the former British ice cover, the more seismically active zones include parts of central and western Scotland but, more noticeably, a swathe that stretches across almost the whole of England, extending into northern France. Here, earthquake activity is significantly more pronounced than in Ireland to the west or the North Sea to the east, with magnitude 5 earthquakes, for example, happening every decade or so, the latest being centred on Market Rasen in Lincolnshire in 2008. Interestingly, the north-western branch of the Rhine Graben, whose offshoots hosted the damaging 14th and 16th century quakes in the Dover Straits mentioned earlier, also extends into this seismically active quadrant, providing the opportunity for stresses associated with stretching across the graben to interact with those arising from glacial rebound, perhaps explaining these events, and establishing conditions for further quakes to come. Elsewhere in Europe, the low level of earthquake activity in northern Germany is explained within the Muir-Wood model by the fact that it nestles within one of the seismically quiet shadow zones associated with the great Fennoscandian Ice Sheet. Further south, a burst of earthquake activity during the early Holocene, together with the rare large historical earthquakes—including the Basel disaster—may be explained as a consequence of their lying in one of the seismically active zones arising from the wasting away of the smaller ice mass that capped the Alpine region. In America, the New Madrid earthquakes, along with that which struck Charleston, South Carolina, in 1886, sit foursquare within Muir-Wood's seismically active quadrant, which occupies a position to the south-east of the long-departed Laurentide Ice Sheet.

There remains much that we don't understand about the precise nature of links between the wholesale melting of the great northern hemisphere ice sheets, the resulting bouncing back of the lithosphere, and the consequent revival of suppressed earthquake activity. In particular, the degree to which current and future earthquake hazard in Europe and North America are legacies of the lost ice continues to tax the brains of geophysicists and seismologists on both continents. The evidence supporting a vigorous seismic response to deglaciation is, however, indisputable and certainly convincing enough to have climate scientists and geologists pondering just how the crust beneath our feet will react if and when the great ice masses that cover Greenland and West Antarctica ultimately go the same way as the ice sheets of the late Pleistocene.

Heavy water

Ice is heavy, but the liquid variety of H_2O is even heavier; the reason, of course, that ice floats on water. While a cubic metre of water weighs a metric tonne, an equivalent volume of glacier ice turns the scales at a mere 850 kilos or so. It should not be surprising to discover, therefore, that dumping large quantities of water on the Earth's crust, or taking it away, has a similar effect to loading or unloading with its solid counterpart, and volume for volume the liquid version is even more effective at eliciting a response, as detailed later. It is also far more adept at insinuating itself deep into the crust itself, a particularly effective means of promoting a seismic reaction. Times of abrupt climate change involve huge variations, not only in the quantity of ice draped across the top and bottom of the world and in high mountain regions, but also in the volume of water contained in our planet's lakes, seas, and oceans. Inevitably, roller-coaster sea-level changes that matched the height of a 30-storey building and the growth and emptying of prodigious lakes of glacial meltwater left their mark on the crust beneath, and there is plenty of evidence for such behaviour driving fault movements and earthquakes. However, in order to get a decent picture of how changing water mass can shake up the underlying crust, we don't have to go back all the way to the last ice age; half a century or so will be perfectly sufficient.

Water—either too much of it or too little—has always been a problem in South Asia, a predicament revealed in 2010 and 2011 by epic floods that drowned much of Pakistan's Indus valley and affected more than 20 million people, or by the torrential monsoon rains that inundated the Indian city of Mumbai five years earlier and took hundreds of lives, or, further back in time, by the unforgiving and widespread droughts that wiped out millions in late 19th century India. Across much of the subcontinent a battle, the origin of which is lost in the mists of antiquity, continues to be waged with water; on the one hand it is aimed at keeping it out of towns and cities, and on the other, keeping it in wells, ponds and reservoirs so as to provide a ready supply when the rains fail.

It was with this second goal in mind that a plan was developed in the early 1960s to impound the Koyna river in the west Indian state of Maharashtra, where it occupies a deep valley 200 km south of Mumbai. Not only would this scheme ensure that the state's water supply was more secure, but it would have the added bonus of generating cheap hydroelectric power in a part of the country where the wood fire and the oil lamp remained the primary means whereby meals were cooked and homes lit. Construction of the Koyna Dam began in 1962, and by the following year, the river found its natural course barred by an enormous concrete structure more than 100 m high and stretching for 800 m across the valley. Water rapidly accumulated behind the obstruction, forming the 50 km long Lake Shiva-jisagar and filling the valley to a depth just below the rim of the dam. Electricity soon began to flow into the surrounding countryside and everyone was happy—for a while.

At the time of construction, the location of the dam and reservoir was regarded as geologically quiet and the area certainly had no particular reputation for earthquakes. Imagine the surprise and concern of the authorities and the local population when, after the filling of the reservoir, the Earth started to tremble. The frequency and size of tremors increased from 1963, building towards jolts exceeding magnitude 3 in November and early December 1967. Still tiny in comparison to the cataclysmic shocks that terrorize populations living close to the margins of the Earth's tectonic plates, but large enough to lead to disquiet in an area where seismic activity was very rare. In fact, the worries of the local population were perfectly justified, as a little over a week later, at around four in the morning, the area was rocked by a far more serious magnitude 6.3 earthquake. The 11 December quake was located close to the site of the dam, which took a severe battering. Fortunately the structure held, otherwise close to three billion cubic metres of water would have erupted into the valley below, condemning settlements downriver to near-instant annihilation. As it was, the quake still proved to be lethal, resulting in more than 200 deaths and perhaps as many as 2000 injured, mainly due to collapsing buildings. This toll is very small for an earthquake of this size occurring in a poorly developed nation in the 1960s, but is a reflection of the low population density rather than well-built homes. In Koyna Nagar, closest to the epicentre, around 80 per cent of the buildings were flattened, while in neighbouring villages no buildings are reported to have survived at all. Bridges and roads were cut by fissures, rubble, and landslides, but the power station within the dam survived reasonably intact, although the lights went out across the region, and as far as Mumbai, and production at textile mills in Mumbai and Poona was interrupted.

Storing up trouble

While the dead were buried, the streets cleared of debris and the dam and power station repaired, geophysicists and seismologists pondered long and hard upon the nature of the earthquake and its cause. The coincidence in timing between the impounding of Shivajisagar Lake and the onset of tremors in this seismically quiet part of the subcontinent was too much to ignore, and it was not long before scientists, the media and the local people were blaming construction of the dam for stirring up the underlying crust.

The idea that new reservoirs were capable not only of storing water but also of storing up seismic trouble for the future was not new, even in the mid 1960s. The first report of the weight of water impounded behind a dam inducing earthquake activity came from the USA in the 1930s. Construction of the mammoth Hoover Dam across the Colorado River began in 1931 and was completed five years later. The 220 m high dam formed an elegant concrete arch that joined Arizona in the west to Nevada in the east, backing up the river waters to form Lake Mead. Still the largest reservoir in the USA,

Lake Mead holds more than 35 cubic kilometres of water and has a surface area of close to 650 square kilometres; that's about four times the area of the national capital, Washington DC. Like Koyna in India, the area of the south-western USA where the great dam was built had no history of earthquake activity, until, that is, the waters began to accumulate behind it. More than 600 earthquakes rattled the dam and the surrounding area as the impounded waters rose towards a peak of 150 m—almost exactly the same height as the Statue of Liberty—which was reached in 1939. Fortunately, for the survival of the dam and for local people, no quake bigger than magnitude 6 was recorded, with the greatest shaking arising from a single magnitude 5 event. This is still a pretty impressive response, however, for a piece of the Earth's crust that had previously been seismically subdued.

Over time it has become apparent that the seismic reaction to the reservoirs captured behind the Koyna and Hoover dams is nothing special and such a response from the underlying crust has now been identified at more than 100 locations where dams have been constructed to hold back large volumes of water. Damaging earthquakes were recorded in the late 1950s during construction of the great Kariba Dam, which straddles the Zambezi river between Zimbabwe and Zambia. At the time, the reservoir that accumulated behind the dam was the world's largest; containing something like 180 billion metric tonnes of water. In light of what happened on the Colorado a quarter of a century earlier, it should not have been surprising that filling of the lake should be accompanied by strong seismic activity, including more than 20 earthquakes larger than magnitude 5. Elsewhere, in 1962, serious cracks were opened in China's Hsingfengkiang Dam by a magnitude 6 earthquake promoted by the weight of the accumulating water at its rear, while four years later, a sequence of hundreds of earthquakes linked to the impounding of water behind the Kremasta reservoir in northern Greece culminated in a shock of close to magnitude 6.

Piling on the pressure

It is unlikely to have escaped your attention that while reservoirs appear to be effective at triggering earthquakes, they seem to be doing so in the opposite sense to that arising from glacial activity. At the close of the last ice age, seismic activity on rejuvenated faults is explained as a consequence of the removal of the enormous weight of the overlying ice sheets. Behind the dozens of giant dams constructed in the past half century or more, in contrast, it is the addition of the water's weight that is being charged with triggering earthquake activity. How can this be so? It seems intuitive that removing a great weight from a fault may allow it to metaphorically shake itself down and move more easily, but weighing a fault down with billions of tonnes of water would surely have the opposite effect? Careful analysis of the locations of earthquakes associated with the filling of reservoirs behind dams confirms, in fact, that the increased seismic activity is not linked to the weight of the water at all. Beneath the deepest part of a reservoir, where the load caused by the overlying water is greatest, there is no seismic response, with most earthquakes crowding around the reservoir's margins. This tells us that, in a manner comparable to the effects of the great Pleistocene ice sheets, the weight of the water is acting to stabilize any faults that lie directly beneath the reservoir, while those at the periphery are destabilized and encouraged to move. This idea is further supported by observations of earthquake activity in the vicinity of the Tarbela reservoir, located on the Indus River in the Pakistan Himalayas. Here, small-scale earthquake activity on faults beneath the reservoir seems to be more common during the dry season when water levels are lower, the logical explanation being that underlying faults are actually stabilized by the water load when the lake is full.

How then do reservoirs trigger earthquakes? Unlike its solid counterpart, water is sneaky stuff that can and will insinuate itself into every nook and cranny, and this is the key. Once dam construction is complete and a reservoir starts to fill, water will be forced into the adjacent rock, pushing along fractures both small and large. The infiltration of new water into the rock mass surrounding the new lake will raise the pressure of water already contained in the pores of the rock causing it to exert extra force that weakens the rock. Where a fault is present that is in a suitably critical state, the increased fluid pressure can trigger movement and cause an earthquake by reducing the stresses that have—up to that point—kept the fault stable, In simple terms, the effect of infiltrating water and raised pore pressures is to lubricate the fault, thereby promoting increased ease of movement.

The relationship between changing pore pressures and seismic activity in the vicinity of reservoirs is sometimes so close as to be able to recognize a pattern, as the level of the reservoir changes in response to variations in input and output. At Koyna, for example, elevated levels of earthquake activity are still recorded, more than 40 years after the construction of the dam and the impounding of Lake Shivajisagar. The University of South Carolina's Pradeep Talwani, a pioneer of the study of what has become known as 'reservoir induced seismicity' or RIS, observes that the occurrence of earthquakes at Koyna is far from random and follows a recognisable pattern. The lake level rises during the rainy season, between June and August, and falls steadily over the rest of the year. This picture is matched by that of the local seismicity, which follows on immediately from the rainy season and high lake levels, with the largest earthquakes recorded six to eight weeks after the level of the lake starts to rise. The link between the two provides one of the most convincing pieces of evidence in support of large water bodies triggering earthquakes; the lag between the two is easily explained by the time required for rising pore pressures in the rock surrounding the reservoir to reach values high enough to promote fault movement. Causal relationships between changing reservoir levels and earthquake activity in the vicinity are, in fact, commonly recognized, and have been observed at both Lake Mead, held behind the Hoover Dam, and at the Lake Monticello reservoir in South Carolina.

As more and bigger dams are constructed in a desperate race to meet the 30 per cent increase in global water demand predicted by 2030, along with the world's ever-growing energy needs, so worries over the influence of the newly impounded reservoirs on the stability of the surrounding crust are growing. These came to something of a head in 2008, following the magnitude 7.9 earthquake that obliterated Wenchuan County in China's Sichuan Province, taking close to 70,000 lives and leaving more than four million homeless. Unlike India, large and destructive earthquakes are a part of the landscape in China, which has hosted some of the most lethal and devastating seismic catastrophes ever recorded. Although Wenchuan had been shaken more than half a dozen times prior to the 2008 'big one', the rate at which seismic strain accumulates in the area is small and no major quake had been identified during Holocene times. Nevertheless, it was not long before fingers of blame were being pointed at the newly constructed Zipingpu dam, located just 21 km from the epicentre of the quake, and opened only four years earlier. Could the 120 m deep lake impounded behind the dam have played a role in triggering the earthquake? It certainly made a useful scapegoat for those who had lost everything in one of the most devastating earthquakes in the early years of the 21st century. Their case was supported too by the research of Columbia University's Christian Klose, who presented findings at the Fall meeting of the American Geophysical Union seven months after the event that suggested the earthquake could indeed have been triggered as a consequence of the 300 million tonnes of water impounded behind the new dam. Klose's proposal—that the resulting stress alterations in the crust may have brought closer to failure a fault that was already in a critical state—remains, however, hotly contested. Other researchers have argued that while the reservoir may have had a role in promoting an increase in small, shallow, local shocks, neither the additional water load nor the increased pore pressures could have played a role in initiating a magnitude 7.9 event at a depth of 20 km below the surface.

With some Chinese scientists supporting Klose's interpretation of events, the argument and discussion continues. Already, however, concerns for similar tragedies in the future have spurred Chinese environmentalists to call for a moratorium on the construction of new dams in the region, pending more detailed analysis of the Zipingpu case and its implications. Inevitably, given its vast dimensions, there has long been speculation that the 40 cubic kilometre reservoir impounded by the Three Gorges Dam in China's Hubei Province, east of Sichuan, could promote a serious seismic response that could even bring down the dam. True to predictions, increasing levels of seismic activity have accompanied the rising water levels in the 600 km long reservoir, which reached 175 m in 2009. Whether this will herald a major quake at some point in the future is the question keeping both residents and the Chinese authorities awake at night.

Given that the role of water in triggering earthquakes by means of forcing up pore pressures in the rock is now well established, it might come as something of a revelation that engineers are, at this very moment, in various parts of the world, pumping water under high pressure deep into the crust via boreholes. They are not deliberately

attempting to promote earthquakes, but are looking to extract geothermal power from the Earth's hot interior. Not unexpectedly, however, earthquakes are what they are getting. You would be right in
thinking that, for historical reasons, the Swiss city of Basel was maybe
not the best place to try out this technique, but that did not prevent
Swiss energy company Geopower going right ahead with its Deep
Heat Mining Project. In 2006, five years after the project's start, and
marking the 650th anniversary of the 1356 catastrophe, the drill team
launched the first full-scale injection of water into the crust, at a
depth of five kilometres. The plan was that the water would open up
fractures in the rock, so increasing its permeability, thereby allowing
the circulation of further water pumped down the well, which would
pick up heat and bring it to the surface via another borehole, producing steam that would drive turbines to provide power for 3000 or so
homes. What actually happened was that the injected water triggered
a magnitude 3.4 earthquake that alarmed the city's inhabitants and
caused damage costing some US$9 million. Other quakes followed
and the project was eventually shut down in 2009. The new method of
Enhanced Geothermal (EG) technology tried out at Basel is also being
looked to in many other parts of the world as a future green energy
alternative to hydrocarbons. For example, in the USA, where some
estimates suggest that EG could help the nation derive up to 30 per
cent of its energy from the planet's geothermal heat, a project was
underway to extract heat from an area of hot rock 150 km north of
San Francisco known as The Geysers. With one eye on what happened in Basel and the other on public anxiety, this project too was
closed down, this time by the US Department of Energy. There
remains great interest in this green technology and its potential to
help reduce global greenhouse gas emissions, but only, and quite reasonably, if it isn't accompanied by the added baggage of damaging
and potentially even destructive earthquakes.

A world awash

The rapid jump in global temperatures that brought the last ice age to a sudden close ensured that a world formerly bitterly cold and bone dry was soon awash with water, filling depleted ocean basins, forming giant glacial lakes and feeding a warmer and wetter climate. As the crust responded to the disappearance of the great continental ice sheets by hosting an increase in earthquake activity as it readjusted itself to the new stress conditions, so a seismic riposte also characterized places where large bodies of water accumulated and drained away during post-glacial times. Probably the best known location for this association is the Lochaber area of the western Scottish Highlands. Here, close to the highland town of Fort William, and in the shadow of Ben Nevis, lie the valleys of Glen Roy and Glen Spean. These iconic relics of the post-glacial world, much visited today by students of geography and Earth science, had long attracted the attention of 19th century natural philosophers and scientists including Charles Darwin, the pre-eminent geologist Charles Lyell, and early pioneers of ice-age research, including geomorphologist Thomas Jamieson and Swiss glaciologist and palaeontologist, Louis Agassiz. The last is remembered as the first person to make a scientific case for an age of ice in our planet's relatively recent history, and his name lives on in the giant meltwater Lake Agassiz, that decanted itself repeatedly, and with remarkable effect, into the North Atlantic of the Late Pleistocene and Holocene.

What makes these glens so interesting is a number of enigmatic horizontal lines that parallel one another and follow the contours of the topography. Easily observable at a distance, on closer inspection they resolve themselves into nicks cut into the hillsides enclosing Glen Roy and adjacent valleys. Familiar to locals for many centuries, the origin of these so-called 'parallel roads of Glen Roy' was the source of much speculation. One legend has it that they were carved from the rock by the Celtic giant Fingal, to facilitate his hunting trips; another was that they were constructed by the ancient kings of Scotland for reasons unknown. In fact, the parallel roads owe their existence to natural processes rather than to supernatural prowess or to the work of long-gone Scottish royalty, but they are no less fascinating for that. Both Darwin and Lyell recognized that the roads were not transport routes but preserved shorelines incised into the valley sides. The two scientists assumed a marine origin, but were somewhat stumped by the fact that the shorelines were substantially above sea-level. They concluded, therefore, that sea level must once have been far higher so that marine waters penetrated further inland to fill the glens around Lochaber. They were wrong. As first suggested by Agassiz in 1840 following a visit to the area, and confirmed a few decades

Fig 15. The so-called Parallel Roads of Glen Roy mark the positions of ancient shorelines, left high and dry when impounded glacial meltwater lakes drained catastrophically around 11,000 or 12,000 years ago.

later by Jamieson, the shorelines are actually nothing to do with the sea, but mark the various levels of glacial meltwater lakes fed by the glaciers that once covered much of Scotland.

Numerous studies of the shorelines and associated deposits over the past 150 years or so have enabled a detailed picture to be painted of the lakes and their evolution. They were both sourced and dammed by ice around 11,500–12,000 years ago, during the Younger Dryas cold snap that saw retreating glaciers in Scotland start to advance once more. In geological terms, the lakes were short lived, lasting not much more than half a millennium before Holocene warming forced the ice to retreat once more, opening up an escape route for the imprisoned water. Based upon painstaking analysis of the ancient lake sediments and shorelines, Adrian Palmer and his colleagues at the University of London's Royal Holloway College have been able to show how the lake levels were forced to rise progressively higher by the advancing ice, before falling in stages as the ice retreated. The falls involved three drainage events during which lake waters would have discharged catastrophically, in the manner of Icelandic jökulhlaups.

The really fascinating thing about the lake sediments is that they show evidence of having been strongly churned up at the time of falling water levels. This was first noticed back in the 1980s by Philip Ringrose, during the course of his postgraduate research, and attributed by him to liquefaction of the soft, water-logged sediment due to earthquake activity. It is perfectly possible and reasonable that the seismic events responsible, which may have been as large as magnitude 6, resulted from the reduced load pressing down on the crust as the ice masses damming the lakes wasted away. Given the apparent timing of this activity, an alternative explanation—also given credence by Ringrose and other researchers since—is that the seismic shaking that disturbed the lake sediments was triggered by sudden reductions in the mass of water resting on the crust as the lakes breached the enclosing ice dams. Within this scenario, and as admirably demonstrated at Pakistan's Tarbela reservoir, high lake levels would have acted to stabilize underlying faults, while the speedy reduction in water mass reduced the load on susceptible structures, allowing them to move more easily.

The influence of sudden lake discharges on earthquake activity is also recognized in North America, on an altogether more impressive scale. Immediately to the east of the Rocky Mountains is a vast region of narrow mountain ranges separated by broad valleys, aligned roughly north to south and known as the Basin and Range Province. The mountain chains and valleys are often bounded by faults, some of which seem to have been particularly active at the end of the last ice age. An explanation for this increase in fault movement that involves unloading due to large-scale ice wastage is a nonstarter because this region was not buried beneath a large ice sheet. A local ice cap did, however, cover the nearby Yellowstone Plateau while, more pertinently, two huge lakes stretched across much of the region. Largest of these was Lake Bonneville, the ancestor of the current Great Salt Lake that lends part of its name to the state capital of Utah. At its acme, the lake swamped an area of more than 52,000 square kilometres, occupying most of what is now Utah and stretching into Nevada and Idaho, making it around the same size as Lake Michigan, the second largest of the Great Lakes. This vast body of water was also, at 350 m, impressively deep, and long-lived. It persisted for nearly 20,000 years, before succumbing eventually to higher temperatures and reduced precipitation. The waters of Lake Bonneville reached their highest levels sometime around 17,000 years ago, resulting in their breaking the shackles of the surrounding topography and draining in an immense flood through Idaho's Red Rock Pass. This cataclysmic event must have made an unbelievable spectacle: a wall of water more than 120 m high and travelling at over 110 km an hour hurtling across the basalt rock of the Snake River Plain, excavating the 180 m deep Snake River Canyon and eventually discharging into the Pacific Ocean far to the west via the Colombia River. At its height, almost a million cubic kilometres of water every second would have been exiting the lake, slashing its volume—in just a couple of weeks—by 5000 cubic kilometres. Unlike the much smaller lakes that filled Glen Roy and adjacent valleys, the waters that filled Lake Bonneville and its neighbour to the west, Lake Lahontan, were not sourced by glaciers. Instead, they owed their origin to a changing climate, with conditions of reduced evaporation and increased rainfall promoting the collection of water in regions that in previous times would have been too dry. Conditions supporting lake formation seem to have occurred many times over the previous three million years, with evidence existing for Lake Bonneville having dozens of predecessors.

As in the Scottish Highlands, there is a clear coincidence between the shrinking of Lakes Bonneville and Lahontan and bursts of earthquake activity on local faults. Andrea Hampel of Germany's Ruhr University, Bochum, has worked with others to unravel the relationship between the great Late Pleistocene lakes of the Basin and Range Province and the activity of local faults, in particular the Wasatch Fault that skirted the eastern margin of Lake Bonneville. This is a major tectonic structure nearly 400 km long that is still very much active, hosting a powerful earthquake every 350 years or so. There is serious concern in the region that the next one could be severely damaging to Salt Lake City and other population centres along its length. The big worry is that more than two million people live and work in buildings constructed on the soft sediments that once made up the floor of Lake Bonneville. These sediments are likely to liquefy readily during strong seismic shaking, leading—according to one estimate—to massive destruction capable of taking more than 6000 lives and causing damage costing $40 billion. During the late Pleistocene, when Lake Bonneville was full, the Wasatch Fault slumbered. The catastrophic reduction in the volume of the lake around 17,000 years ago seems, however, to have provided a wake-up call, and a burst of earthquake activity followed soon after. Hampel and his colleagues determined that this was reflected in parts of the fault slipping, during this period, at twice the normal rate, a direct consequence of the reduction in the load exerted upon them by the shrinking lake. This resulted in the crust rebounding isostatically by up to 70 m, so freeing up the fault and initiating a pulse of earthquakes. A similar, although less marked, response is also detectable on faults in the vicinity of Lake Lahontan, where a smaller volume of water drove a rebound of not more than 20 m.

What's going on in Alaska?

There is plenty of evidence, then, to support a relationship between water—in either its solid or liquid form—and earthquakes; this is coming both from the time of rapid climate transition that marked the latest Pleistocene and early Holocene and, far more recently, from our experiences of dam building and the impounding of artificial reservoirs. The natural changes responsible for encouraging more earthquakes were impressive, involving the wasting away of ice sheets kilometres thick or the spontaneous draining of giant lakes. What would be nice to know is whether processes related to contemporary climate change are also capable of triggering a seismic response. The places to look, in an attempt to answer this question, would be those parts of the world that are experiencing the most rapid rates of ice-mass loss or glacial meltwater accumulation—somewhere, for example, like Alaska. Like Iceland, this detached state of the USA is a land of fire and ice, wherein Pavlof and 42 other historically active volcanoes rub shoulders with snow-covered peaks, vast stretches of permafrost and spectacular glaciers. When one of the state's more lively volcanoes is in full eruption, Alaska can be rather too hot for most people to handle, but most of the time it is cold—very cold. In common with most high-latitude regions, Alaska is not, however, as cold as it used to be. Average winter temperatures have climbed by more than 3°C in the past 50 years, resulting in wholesale thawing of permafrost. Infrastructure is already suffering, with cities like Fairbanks increasingly characterized by misaligned buildings, deformed bridges and fractured road surfaces, as their foundations struggle to maintain an upright stance in the increasingly soggy ground. In the wilds too, the rapidly warming climate is making its mark, especially on the great glaciers. In the south-east of the state, the Bagley Ice Field (sometimes known as the Bagley Ice Valley) still hangs onto its position as the largest, non-polar, ice field in North America, but it is shrinking rapidly. Currently 200 km long, 10 km wide and in places up to 1 km thick, the ice field feeds dozens of smaller valley glaciers that push outwards like fingers from the parent ice mass. These include the Bering Glacier, one of the largest in the USA. Now, though, the valley glaciers, including the Bering, are in full retreat and the ice field itself is beginning to suffer the consequences of rising temperatures and changes in precipitation due to continuing human-induced planetary warming. Since 1900, the Bering Glacier has thinned by several hundred metres and its snout has retreated by 12 km. Periodically, every couple of decades, the retreat is reversed as ice flow increases and the glacier surges. At these times, the snout advances for a short time but a return to relentless retreat follows soon after. So rapidly is the glacier now melting that recent estimates suggest it is releasing into the Pacific Ocean an extraordinary 30 cubic kilometres of water a year—more than twice the flow of the great Colorado River.

Fig 16. The Bering Glacier, one of the largest in the United States, has thinned by several hundred metres since 1900, while its snout has retreated by 12 km. So rapidly is the glacier now melting that recent estimates suggest that it is releasing 30 cubic kilometres of water into the Pacific Ocean every year – more than twice the flow of the great Colorado River.

The surging and retreating glaciers and thawing permafrost testify to dynamic change at the surface, and are phenomena that will become increasingly apparent as anthropogenic climate change hits harder. Beneath the surface too, there is a great deal happening, a legacy not of human activities but of the powerful tectonic forces that come together beneath Alaska and around its margins. The dozens of active volcanoes and the occurrence of some of the world's largest earthquakes, notably the Good Friday Earthquake in 1964, which was the second largest seismic event ever recorded, bear witness to the fact that there are few places on Planet Earth where its geological power is so openly displayed. Beneath southern Alaska, the immense Pacific Plate—at four times the area of North America, the largest of all the tectonic plates—is plunging north-westwards under the North American Plate on which the state sits, at a rate of more than 5 cm a year. This is fast in terms of plate movement and such is the power of the collision between the two great chunks of the Earth's lithosphere that mountain peaks of more than 5000 m have been forced upwards within just 15 km of the Gulf of Alaska. The same tectonic forces ensure that if an earthquake hasn't just happened, one will be along soon, and active faults abound. In addition to the 1964 event, seven other quakes larger than magnitude 8 have shaken the state since the end of the 19th century, four of them causing serious damage. Furthermore, around 6 per cent of all the large, shallow, earthquakes that occur around the Pacific Rim are hosted by Alaska, totalling around 4000 a year.

The combination of fast-vanishing ice and a land sliced through by active faults and racked by earthquakes provides a recipe for some intriguing interactions, which have formed a focus of study for Jeanne Sauber, of NASA's Goddard Space Flight Centre in Maryland, and Bruce Molnia, of the US Geological Survey. They are interested in whether and how continuing wastage of the glaciers and ice fields of southern Alaska influences the stability of faults at the subduction zone plate boundary that immediately underlies the region and, in particular, if shrinkage of the glaciers is reflected at all in the seismic record. Since the end of the 19th century, progressive melting driven by a warming climate has thinned the ice across much of the region by hundreds of metres, and perhaps 1 km in those areas of greatest ice loss. Over this period, while the ice dwindled, the segment of the plate boundary fault deep beneath the surface remained locked, accumulating strain that was eventually released in a magnitude 7.2 earthquake in 1979. The question that begs asking is: did the wholesale loss of the overlying ice mass over the previous century or so trigger the earthquake or at least play some role in its instigation? The answer, according to Sauber and Molnia, would appear to be, maybe. They conclude that shrinkage of the glaciers over the period would certainly have resulted in a large reduction in the stability of underlying faults. This would have amounted to a pressure fall comparable to 20 times that exerted by the Earth's atmosphere and would certainly have been sufficient to facilitate the type of thrust faulting that characterized the 1979 quake. While this does not mean that the ice loss was a direct cause of the earthquake, it does indicate that it would have made it easier for the fault to slip and could have brought forward the timing of the event. In other words, while tectonic forces deep within the Earth provided the driving force, the reduction in ice mass at the surface supplied a helping hand.

Evidence for short-term ice fluctuations in southern Alaska influencing local earthquake activity is also becoming apparent. During the mid 1990s, for example, the Bering Glacier took off on one of its periodic surges, which resulted in the rapid shifting of an enormous 14 cubic kilometres of ice from the Bagley Ice Field source to the glacier snout. According to Sauber, Molnia and their co-workers, the resulting thinning of ice in the source region promoted a noticeable rise in the number of small (less than magnitude 4) earthquakes in the underlying crust, due to the reduced load. More recently too, the number of minor seismic shocks beneath Icy Bay, on the coast immediately to the south of the Bagley Ice Field, increased in the early years of the new millennium, coincident with a significant rise in the rate of ice wastage.

The news from Alaska has great significance from a seismic hazard point of view as it reveals that 10,000 years after the end of the last ice age, changes in ice mass continue to trigger, or at least facilitate, earthquakes. At high latitude locations, such as Alaska, where ice-covered terrain coincides with active faults capable of generating large and destructive earthquakes, and in comparable situations at icebound terrain remote from the polar regions, such as the Andes and New Zealand's Southern Alps, this recognition calls for the seismic hazard to be looked at again. Normally, the evaluation of earthquake hazard for a particular locality is based simply upon its seismic record over time and is unlikely to take into account changes in environmental factors such as ice mass. The sterling work in the wastes of Alaska undertaken by Sauber, Molnia and others committed to learning more about the interactions between ice and earthquakes, makes the point that for glaciated areas, earthquake hazard evaluations should take into account the past behaviour of the ice as well as that of the Earth's crust. And not only the past, but the future too. With prospects for the survival of our world's glaciers seeming increasingly bleak, establishing the likely consequences of their removal for the active faults that lie beneath takes on even greater importance.

Lessons from the past and present

As demonstrated earlier for volcanic activity, it is clear once again that isolating what is going on within the Earth's crust—in this case earthquakes—from what is happening at the surface is not possible. Water, either in its solid or liquid form, is once more shown to be capable of influencing potentially hazardous geological activity and garnering responses from the interior of our world. The main lesson we should take from the post-glacial period and from our recent dam-building campaigns is that the removal or addition of ice and water—in sufficiently large quantities—can have a significant impact on the behaviour of underlying and adjacent faults. Broadly speaking, adding an ice sheet or thickening a glacier will act to stabilize any faults lurking beneath, suppressing their activity and reducing the incidence of earthquakes. Removing an ice sheet, or thinning a glacier will have the opposite effect, freeing up said faults and facilitating a seismic response. Where water is concerned, the loading and unloading effects are comparable, but there is also the added complication that the pressurisation of pore waters in the rocks adjacent to new or growing lakes or reservoirs has also been shown to be effective at destabilising nearby faults.

With climate change driven by human activities very likely, if present trends continue, to bring temperature and sea level rises comparable to those of the post-glacial period, it would be a surprise indeed if at least some of the countless active faults that crisscross our planet did not respond to the new redistribution of global water that is certain to occur as our world continues to heat up. Melting ice sheets at high latitudes, and thinning glaciers in mountainous regions, rapidly rising sea levels, the accumulation and draining of glacial lakes in regions like the Himalaya, and the building of giant dams to tackle predicted water shortages all have the potential to trigger a seismic reaction from the underlying crust. Whether we will be able to detect a noticeable rise in the global earthquake record is debatable, and it is perfectly possible that any increase in earthquake numbers or frequency on a planet-wide scale will be hidden in the statistical 'noise' associated with the couple of million earthquakes that happen, on average, every year. Regionally or locally, however, a response from underlying faults may be more obvious, particularly when and if the great ice sheets that cover Greenland and West Antarctica start to break up. It may well be that the crust beneath these vast icy wastelands is relatively stable and contains little in the way of active faults capable of triggering earthquakes. Alternatively, the ice may be the only thing keeping the slumbering faults beneath from bursting into life. Although the occurrence of even quite large earthquakes in sparsely populated Greenland or penguin-dominated West Antarctica may not seem to present much of a problem, their effects could be felt much further afield. Large earthquakes are especially good at triggering landslides where the crust is unstable, and these in turn, if they occur in a watery environment, can promote tsunamis that can remain destructive and lethal many thousands of kilometres from their sources. As we will examine in the next chapter, during the early part of the Holocene, as Scandinavia continued to shed its icy covering, earthquakes promoted by isostatic rebound are thought to have provided the trigger for massive submarine slides of glacial sediment off the coast of Norway, which sent tsunamis hurtling across the north-east Atlantic to crash into the Shetland Islands and the Scottish mainland. A future burst of earthquake activity in Greenland may not, therefore, seem quite so benign from the perspective of a coastal inhabitant of Iceland or Newfoundland.

5

Earth in Motion

I can still remember my first sight of the Valle del Bove—a gigantic amphitheatre ripped out of the eastern flank of Mount Etna, Europe's largest and most lively volcano. It was May 1977 and I was visiting the volcano for the first time, prior to starting a research project for my PhD. To a fresh-faced lad, recently graduated in geology, the astonishing panorama came as something of a shock. Perched on the edge of a rocky buttress, with the summit craters of Etna grumbling softly behind me, I looked into a truly gigantic hole. Vertical cliffs of lava, relieved only by vertiginous slopes of ash and debris, plunged down a kilometre to the floor of the chasm, upon which flows of solidified lava wound this way and that amid boulder-strewn scrub and trees diminished by distance to toy-land dimensions. To left and right, the perimeter cliff walls of the amphitheatre stretched far into the misty distance, before curving forwards

161

to a remote opening some 8 km away. Here the land fell away to a vividly blue Mediterranean that twinkled invitingly in the early morning sun.

The Valley of the Oxen

To the constant stream of tourists transported by jeep and off-road bus to the viewpoint throughout the year, the extraordinary Valley of the Oxen—as its name translates—meant just one more stop on a packed itinerary. An unexpectedly sublime one, true, but still just another ticked box; another photo for the album. For me, that first glimpse marked something very different—the beginning of a research career that has since enabled me to spend three decades and more exploring dead, dozing and dangerously awake volcanoes across the planet. What would prove to be the first step on this continuing tour, however, was three years spent attempting to unravel the secrets of this astounding but poorly understood topographic anomaly, through studying and sampling the volcanic layers exposed in the surrounding cliffs, trying to build a picture of the great amphitheatre's origin and the circumstances that conspired towards the excavation of around 12 cubic kilometres of rock from the eastern flank of the volcano some-times known as Vulcan's Forge. More than three decades on, my rudimentary analysis and the more exhaustive work of others since, has enabled a picture to be painted of how the Valle del Bove probably came into being.

Recent dating campaigns suggest that the amphitheatre had a catastrophically violent birth around 7500 years ago, during the early Holocene, that post-glacial period when the Earth environment was still getting used to its new ice-free (comparatively speaking) status. In keeping with many parts of the world, the climate of Etna around this time had switched from arid and icy to wet and warm. Persistent heavy rainfall sliced through the volcano's rocky carapace forming yawning valleys, and raised water tables so that the giant edifice of ash and lava became saturated. Torrential rains transformed the volcano's ashy mantle into raging mudflows that poured periodically down its flanks while an explosive conjugation of water and magma blasted out columns of rock, debris, and dust. Deep within the volcano more magma forced its way eastwards from beneath its summit, heating the water trapped within the pores of the sodden rock and compelling it to expand. Spontaneously and catastrophically, the rising pressures exerted by the boiling ground water caused the volcano to break in two. In a matter of seconds, a gigantic chunk of its eastern flank detached itself and careered downslope faster than a Formula One racing car, tumbling and disintegrating as it went. Crashing into the Ionian Sea, the immense mass of volcanic debris— equivalent in volume to around 3000 Wembley soccer stadiums— gave birth to a tsunami that hurtled eastwards. Based upon computer modelling, Italian volcanologist, Maria Pareschi and her co-researchers reckon that the tsunami would have affected the whole of the eastern Mediterranean, including Greece, the Levant, and the coast of North Africa. If their model is accurate, the tsunami run-up may have been as high as 40 m on the Calabrian coast of southern Italy—about the height of a 12-storey building—falling to 8–13 m in Greece and Libya and 2–4 m on the Egyptian and Levantine coasts. The Italian research group suggests that the tsunami might have been responsible for the destruction of the Neolithic village of Atlit-Yarn on the coast of Israel, which, archaeological excavations suggest, seems to have been overrun by a tsunami at around this time.

Such an event occurring in our modern world would be cataclysmic, not only for those living on the densely populated flanks of Mount Etna and the crowded coastline of eastern Sicily, but also for the Mediterranean in general. Collapsing volcanoes are far from rare

Fig 17. The Valle del Bove is a vast amphitheatre carved from the eastern flank of Mount Etna in Sicily. The excavation of this great cavity is attributed to the onset of a warmer, wetter, climate, which triggered the formation of a giant land-slide around 7500 years ago.

and have taken many lives in the last few centuries alone. As I will explain in due course, like Etna's Valle del Bove, they also demonstrate a worrying proclivity for crumbling, during times of dramatic climate transition.

Nature's rubble piles

Anyone would be forgiven for thinking of giant volcanoes such as Mount Etna, Japan's Mount Fuji, and the great Hawaiian volcano, Mauna Loa—the largest mountain on Earth—as timeless sentinels: bastions of strength and rigidity that are unmoving and unmovable. This perspective, however, is well off the mark and even the mightiest volcanoes are morphologically dynamic structures that are constantly shifting and changing as fresh magma insinuates itself into the interior or pours or explodes out at the surface. Far from being robust, they are more often than not rotten to the core; little more than wobbly piles of ash and lava rubble looking for an excuse to come crashing down. Because of constantly circulating hot fluids, the hearts of many volcanoes have also been weakened severely by hydrothermal weathering, turning much of the interior to a clayey mush that is far easier to destabilize than solid rock. It is somewhat indicative that many laboratory experiments designed to emulate the behaviour of volcanic edifices make use of the material gelatin or—as it is more commonly known in its fruit-flavoured form—jelly.

In addition to their unsound make-up, there are many reasons why active volcanoes quickly become unstable and prone to sideways, or lateral, collapse. Uplift, subsidence or tilting of the rocky substratum upon which a volcano has grown and developed may promote instability over a long period of time, as may environmental factors such as rising or falling sea levels—especially at times of abrupt climate change. Because most volcanoes are located at or close to the margins of the tectonic plates, where earthquakes are also common, a ready trigger is always available to help a volcano rid itself of a detachable flank or part thereof. The fact that many volcanoes form strongly elevated terrain means that they are often either glaciated or have local climates that are wet. As became apparent in the previous chapter, water in either its solid or liquid form plays a role in earthquake triggering, which can, on its own, persuade a volcano to collapse. In addition, heavy rainfall and snow and ice melt can also promote the formation of landslides, particularly through their pressurising effects on pore waters contained within the hearts of volcanic edifices, a key reason why volcano instability and collapse, on the one hand, and climate change, on the other, appear to be closely linked.

Even a cursory inspection of the data will reveal that, broadly speaking, the biggest collapses occur at the largest volcanoes, which have been active for many thousands, if not millions, of years. The biggest of all have periodically removed huge chunks of ocean island volcanoes, such as those making up the Canary Island and Hawaiian archipelagos, leaving behind gigantic amphitheatres carved out of the solid rock. Nowhere is the legacy of collapse better exhibited than on the Canary island of El Hierro, where giant collapses in prehistoric times have sculpted the island into a shape approaching that of a tricorne hat. In total, the seven islands that make up the archipelago host evidence for at least a dozen great collapses, revealed either by the obvious existence, on land, of the scars they have left behind, or by the trains of disrupted debris imaged by sea-bed surveys. On the popular holiday island of Tenerife, the catastrophic removal of massive amounts of rock more than half a million years ago excavated the spectacular Orotava and Icod valleys, while on neighbouring Gran Canaria, the 80 m high iconic monolith known as the Roque Nublo is a surviving remnant of a giant collapse that removed much of the southern flank of an ancient volcano around three million years ago. Next door, on the island of La Palma, the vast hole of the Taburiente caldera, gouged out by a landslide more than half a million years ago, dominates the northern part of the island. It is bounded by vertiginous cliffs more than a kilometre high, and is so large that it takes two days to walk into. South of the Taburiente collapse is the Cumbre Vieja volcano, the most active in the archipelago during recent times and the focus of enormous interest by geologists and geophysicists. This is mainly because, during an eruption in 1949, a large chunk of the western flank of the volcano dropped by a few metres towards the sea, suggesting that it was becoming increasingly unstable. Arguments within the scientific community and beyond about when and if this volcano will collapse into the North Atlantic, and the scale and extent of the tsunami that may result, have been vociferous and sometimes downright vitriolic, but more about this later in the chapter.

Fig 18. The Canary Island of El Hierro; sculpted by giant collapses in prehistoric times into a shape approaching that of a tricorn hat.

The largest of the great volcano collapses of the Canary Islands have volumes of a few hundred cubic kilometres—colossal when you think that the awe-inspiring failure of the north flank of the Mount St Helens volcano in 1980 involved just 2.5 cubic kilometres of rock and debris. When compared to the sizes of the landslides that have removed substantial parts of the Hawaiian Islands, however, they pale into insignificance. There is clear evidence for more than 70 gargantuan collapses from the flanks of the volcanoes that make up the Hawaiian archipelago, the largest of these having volumes in excess of 1000 cubic kilometres. The grand-daddy of them all is the Nuuanu landslide, which is spread across the submarine flank of the (probably) extinct Koolau volcano that makes up the eastern half of Oahu—home to Honolulu. Formed a few million years ago, the landslide took a massive bite out of the flanks of the aforementioned Koolau, and is now scattered across a huge 23,000 square kilometres of the adjacent sea bed, an area about as big as the US state of Vermont, and just a little less than the size of the UK principality of Wales.

The volume of the landslide, at 5000 cubic kilometres or more, is even more mind-numbing, and broadly comparable to what you would get from excavating New York's Long Island, plus a fair bit of the surrounding sea bed, down to a kilometre or so. The sizes of individual blocks in the landslide deposit are also astonishing. One, known as the Tuscaloosa seamount, is 30 km long by 17 wide, and towers almost 2 km above the floor of the Pacific. The Nuuanu landslide travelled around 230 km before it came to a stop and, in common with other giant volcanic landslides, it moved at break-neck

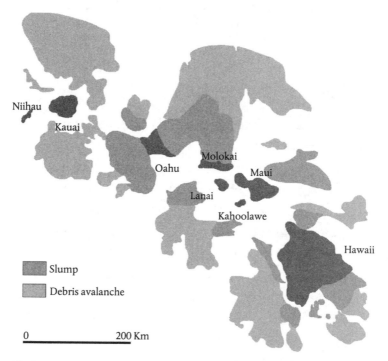

Fig 19. Submarine imaging reveals the existence of more than 70 gargantuan collapses from the flanks of the volcanoes that make up the Hawaiian archipelago, the largest of which have volumes in excess of 1000 cubic kilometres.

speed. Once it got going, the mass of debris probably accelerated rapidly to a speed of 270 km per hour, and possibly more, a speed that would not embarrass even the best Formula One driver. Such was the colossal amount of energy generated by the collapse that, after crossing the sea-floor depression known as the Hawaiian Trough, a 5 km deep 'moat' surrounding the Hawaiian Islands and caused by their crushing weight, there was still sufficient to drive the moving mass for a distance of nearly 150 km up the other side. Imagine the scale of the resulting tsunami!

Mount St Helens...and more

While far, far smaller than their marine counterparts, volcano collapses on land have the advantage, for geologists, that they are much more common and can be seen happening—if we are looking. Provided that we are not too close, we may even survive the experience. Because the volcanoes tend to be smaller, so do the sizes of any landslides arising from their break-up, so rather than hundreds or thousands of cubic kilometres, we are talking here about collapse volumes of anything from a few to a few tens of cubic kilometres—in other words, a thousand or so times smaller than the greatest of the Hawaiian landslides. Arguably the biggest careered down the side of Mount Shasta—a snow-covered neighbour of Mount St Helens in the western United States—around a third of a million years ago, transporting 45 cubic kilometres of volcanic debris to a distance of 50 km. Equally impressive was another prehistoric collapse that took place about 18,000 years ago and which took a 25 cubic kilometre chunk of rock out of the side of Mexico's Nevado de Colima volcano, shifting it, in pieces, into the Pacific Ocean 120 km to the west. Other impressive collapses from volcanoes that have their feet firmly planted on land have been identified at Fuego and Pacaya in Guatemala, Taranaki (sometimes known as Egmont) in New Zealand and, of course, at Italy's Mount Etna.

In the 20th century, 1956 saw Bezymianny, just one of a nest of volcanoes occupying the frigid wastes of Russia's Kamchatka Peninsula, shed its eastern flank, the sudden reduction in pressure on the magma beneath triggering a devastating sideways blast and a violent eruption that continued for four hours. Fortunately, the area is unpopulated so no lives were lost on this occasion. This was not the case, however, when Mount St Helens volcano followed suit in a carbon-copy event, a quarter of a century later. Like its eastern counterpart, Mount St Helens had been quiet for some time, with the result that the vent at the summit of the volcano was blocked. In March 1980, after more than 120 years of lying low, the advent of swarms of small earthquakes announced the arrival of new magma within the volcano—one of 13 volcanic peaks in the Cascade Range that stretches from the Canadian border to northern California. With no possibility of escape via the most obvious route, the fresh magma forced its way into the north flank of the volcano, causing it eventually to bulge outwards by more than 150 m. By early May, well aware of the goings-on in Kamchatka, just 24 years earlier, monitoring scientists were warning of the likelihood of the collapse of the north flank and the probability that this would trigger an explosive blast. No one, however, was quite prepared for the scale of the event, when their prophecy came true on 18 May. At just after eight in the morning, a magnitude 5.1 earthquake immediately beneath the north flank promoted its detachment and sent it hurtling downslope at speeds of up to 250 km an hour. A shock wave arising from the explosive release of gas from the suddenly exposed magma body and travelling perhaps as fast as the speed of sound for a time, rapidly overtook the landslide, obliterating a 600 square kilometre area of mature forest, and reaching nearly 30 km from the volcano. This lateral blast also took all 57 lives claimed by the eruption, including that of geologist David Johnston, who was on monitoring duty to the north of the volcano. Not surprisingly, the removal of the north flank of Mount St Helens, together with its summit, finally caused it to blow, blasting a column of ash to a height of 25 km in just 15 minutes and depositing ash across 11 states. The eruption was the largest in the contiguous United States since Lassen Peak in California in 1915, and ultimately cost the national economy more than one billion dollars.

The collapse of Mount St Helens is certainly the best studied, and undoubtedly it marks the beginning of acute scientific interest in the stability of volcanoes, and the dangers presented by the failure of volcanic edifices and the formation of volcanic landslides. Partly as a consequence of what happened at Mount St Helens, therefore, we now know that such events are far more common than previously thought. The last three and a half centuries alone have seen eight lethal volcanic landslides that, together, have taken close to 25,000 lives, but the average number of collapses per century over the last half a millennium may be as high as 20. A recent survey has identified, in all, 480 collapses at 316 volcanoes, a figure that—based upon our knowledge of the frequencies of such events from the historical record—is certain to be far below the true total. Acknowledging the fact that a hundred million people now live in the vicinity of a volcano that has collapsed in the past, and bearing in mind that such collapses appear to have some preference for occurring at times of climate transition, this is clearly a threat that is not going to go away.

The climate dimension

By now, it should be apparent that there is a clear climate dimension to hazardous geological phenomena, with dramatic changes to the environment heavily implicated in the incidence of both earthquakes and volcanic eruptions. Landslides—volcanic or otherwise—also demonstrate a particular appetite for times when temperatures, rainfall, and other factors that have the potential to destabilize large masses of rock are in transition. As far as volcanoes are concerned, the idea that instability and collapse are influenced or modulated by climatic change is always going to be difficult to prove irrefutably, partly because of the incomplete record and partly due to errors in the dating of some collapses, which make direct correlation with a changing climate problematic.

A picture is beginning to emerge, however, largely as a consequence of the construction, at University College London, of a database of all known volcano collapse events. Of the close to 500 collapses recognized in the geological record, most occurred in the late Pleistocene or Holocene and have ages of less than 40,000 years. As might be expected, more and more collapses come to light as we approach the present day; mainly a reflection of the fact that younger collapses will be less prone to erosion or being covered by later volcanic products. As a consequence, almost half of all dated volcano collapses happened in the past 10,000 years, during the Holocene Epoch, while progressively fewer have been identified further back in time. In order to search for a link with climate change, it is desirable to try to reduce this age bias in the record. One way of doing this is to ignore all the recent collapse events— say those occurring within the past 2000 years—and all those older than 40,000 years, which, conveniently, just about marks the useful limit of the powerful carbon-14 dating method. In addition, setting aside all collapses with volumes of less than one cubic kilometre will also help reduce any bias arising from the improved chances of preservation of smaller collapses of younger age. While those that remain still provide nothing like a complete record of collapse over the period, they do enable us to have a better chance of teasing out variations in the collapse rate that may hint at some sort of link with climate change. The period from 40,000 years to 2000 years ago was, of course, a time of climate extremes, encompassing the Last Glacial Maximum, when the ice sheets reached their maximum extent, and the Holocene, during which the world saw itself transformed into its present warmer and wetter state. If there is any connection between a changing climate and the incidence of collapsing volcanoes, then it surely must show up over this time span, even if the record is incomplete. According to UCL's Rachel Lowe, it does. Lowe looked at the distribution of volcano collapses over the period and noticed that their distribution in time is far from consistent. Most obviously, since the last glacial maximum, a number of distinct troughs and peaks are seen in the record that coincide, respectively, with cold snaps and succeeding warmer episodes. For example, one trough in the rate of volcano collapses coincides with the return to cold and arid conditions marked, a little less than 13,000 years ago, by the onset of the Younger Dryas, and another with the 8.2 ka cooling event.

In sharp contrast, the return to warmer conditions that followed each is marked by peaks in the collapse record. Other less extreme switches from warm and wet to cold and dry and back again, between 6000 and 2000 years ago, also appear to be reflected in the collapse record in a similar way. Why should it be that volcanoes seem to shy away from collapse when the climate is cold and dry, but favour collapse when conditions are wetter and more humid? As suggested in my own research and discussed in chapter 3, sea-level change might be a candidate, with rising sea levels driven by a warming climate promoting collapse of the seaward-facing flanks of coastal or island volcanoes by bending the crust beneath them. There is little suggestion, however, that significant changes in global sea level accompanied the rapid switches from cold to warm that seem to be influencing collapse behaviour in the Holocene and latest Pleistocene. For a far more likely explanation we should, perhaps, look to the changes in the availability of water and the loss of ice mass at cold-to-warm climate transitions. As mentioned earlier, removal of ice from a glaciated volcano can be a very effective means of debuttressing the edifice, thereby increasing flank instability and promoting collapse. Increased runoff from melting snow and ice, in combination with a wetter climate, can also act to raise pore-water pressures within volcanic edifices, again contributing, as in the case of Mount Etna's Valle del Bove, to a situation within which sideways collapse is favoured.

The British geologist Simon Day, through his work on the possible future collapse of La Palma's Cumbre Vieja volcano, is also convinced that volcanoes are more likely to collapse when the climate is muggy and damp rather than cold and dry. Day has noticed that the greatest collapses at ocean island volcanoes over the past couple of hundred thousand years—not only in Hawaii and the Canary Islands, but also in the Cape Verde Islands off the west coast of Africa, and on Réunion Island in the Indian Ocean—can be correlated with the sea-surface temperature record. As far as can be ascertained, bearing in mind the errors involved in their dating, major collapses over this period seem to occur when the sea surface temperatures are high rather than low. It is not suggested that the temperature of the sea provides a direct driver of the collapses, but that it reflects other environmental conditions that can and do promote the development of volcano instability and edifice failure. As might be expected, sea-surface temperatures are an indicator of the general global climate, so during glacial times the oceans are relatively cool, whereas during interglacial periods they warm up considerably. At such times, sea-level is high and the world is broadly clement and moist compared with the intervening periods when the ice sheets are dominant. At low latitudes, where the largest oceanic islands reside, Day observes that the warmer seas resulted in changes to the patterns and peculiarities of the Trade Winds—the prevailing easterlies in the tropics upon which global commerce depended in the age of sail. As a consequence, the wind systems dumped significantly greater amounts of rainfall on the upper flanks of any oceanic island volcanoes that happened to get in the way of their circumnavigation of the globe, than in the cooler and more arid times prevalent during glacial periods. According to Day, this would have had the effect of pushing up water tables by several hundred metres, increasing opportunities for rising magma to heat up and pressurize pore waters in the cores of the volcanoes and laying the groundwork for flank collapse.

Day's ideas are supported by Gary McMurtry of the University of Hawaii's prestigious School of Ocean and Earth Science and Technology (SOEST) and collaborating scientists, whose attentions in the 1990s were centred upon the colossal Alika-1 and Alika-2 landslides generated by lateral collapses of the world's biggest active volcano—Mauna Loa—on the island of Hawaii, both of which occurred during interglacial periods when the world was warm and wet and sea levels were high—just as they are today. There doesn't seem to be too much doubt that our current climate is conducive to promoting the lateral collapse of volcanoes, and is likely to become more so. This does not mean, however, that under different climate conditions, volcanoes are completely stable and unmoving. Some of the older collapses in the Canary Islands, for example, appear to have occurred during colder and drier times when sea levels were dropping rapidly, a correlation that is not entirely surprising given the reduced buttressing effect that would result from the relatively sudden removal of huge volumes of ocean water from a volcano's flanks. Broadly speaking, however, it looks increasingly as if collapsing volcanoes have a clear preference for warmth and humidity.

Wave goodbye!

As Gary McMurtry and his colleagues have cautioned in writing, if increased precipitation is the primary driver of collapsing volcanoes then we need to sit up and take notice. Looking ahead to an anthropogenically warmed world, many regions that contain volcanoes that have shown themselves, in the past, to be prone to collapse are slated to get wetter generally or to experience a rise in the frequency of more extreme rainfall events. Where these volcanoes form islands or are coastally located, the big worry is that future collapses will drive tsunamis large enough to transmit death and destruction far from the source—in the worst cases, perhaps even across entire ocean basins. We are used to associating tsunamis with massive undersea earthquakes, such as that triggered by the tearing apart of a 1200 km section of the Sunda Megathrust in 2004, and the huge megathrust quake that struck off the northeast coast of Japan seven years later. Volcanoes, too, are adept at displacing huge volumes of water, and with comparably devastating consequences. In 1883, the cataclysmic eruption of Krakatoa—until then a pretty innocuous volcano nestling quietly between the Indonesian islands of Sumatra and Java—generated a tsunami as high as a 10-storey building that obliterated countless coastal villages and took more than 30,000 lives. Eruptions are not always required to trigger a tsunami, however, and of the 12 volcanic tsunamis known to have occurred in the past 400 years or so, the majority were instigated by volcanic landslides. These include, just five years after the Krakatoa catastrophe, the collapse into the Bismarck Sea of a substantial chunk of the Ritter Island volcano, located 100 km to the north of New Guinea. This largest island volcano collapse during historic time—twice the size of the Mount St Helens landslide—launched a tsunami that swept the coasts of

New Britain and adjacent islands to a height of 20 m, taking an estimated 3000 lives.

Given the extreme violence of volcano collapse events and the staggering amounts of energy involved in the formation and transport of even the smallest volcanic landslide, their tsunami-forming potential should not be a surprise—neither should it be underrated. The Mount St Helens collapse, tiny by comparison with the greatest volcanic landslide events, generated almost as much energy as 2000 Hiroshima atomic bombs. The great Hawaiian slides of the past would have dwarfed this, pumping out as much energy as even the largest and most destructive earthquake, more than enough to power a city the size of Los Angeles for a full year. The transfer of this energy to the water into which an island or coastal volcano collapses is what drives the resulting tsunami and provides it with the capacity to transmit devastation far and wide. The heights reached by past tsunamis attributed to collapsing volcanoes are sometimes hard to credit or visualize. Gary McMurtry, working with the British Geological Survey's Dave Tappin and others, has tracked down debris left behind by a tsunami at elevations of more than 60 m above sea level on the flanks of Hawaii's Kohala volcano. A tsunami powerful enough to hurl itself this high up the flanks of a volcano must have been an awesome sight, but this is only a part of the picture. For the past half a million years or so, Kohala volcano, along with the rest of Hawaii's Big Island, has been sinking under its own weight, so much so that at the time of the tsunami's formation the volcano was sitting close to 350 m higher than it is today. The waves that crashed into Kohala around 120,000 years ago didn't just travel 60 m up the flanks, but reached to the incredible height of more than 400 m. This almost unimaginable wave or, more probably, series of waves, was the legacy of the Alika-2 collapse mentioned earlier, which almost instantaneously removed around 500 cubic kilometres of neighbouring Mauna

Loa volcano and dumped it into the Pacific. Kohala's tsunami deposits are matched by others on a number of islands in the archipelago and mirrored by deposits in the Canaries. Here, on the island of Gran Canaria—well known to Britons and northern Europeans alike for its sunny climate and liquid hospitality—layers of cobbles and shell material can be found more than 180 m above sea level, testimony to a huge, prehistoric volcano collapse in the archipelago, maybe originating at Mount Teide on neighbouring Tenerife, or possibly further afield. Given the large number of collapses in the Canaries, along the Hawaiian archipelago, at the Cape Verde volcanoes and elsewhere, for which there is incontrovertible evidence, giant tsunamis may well have left their tell-tale signatures elsewhere. It could be that some of these have been mapped and interpreted as having been formed by some other geological process. The likelihood is, however, that we have just not searched hard enough or in the right places.

Looking to the future, is there an unstable volcano poised to collapse that might be amenable to a helping hand from climate change? The obvious candidate is La Palma's Cumbre Vieja volcano, whose story caused such uproar within the tsunami community following its exposure on the television documentary—*Megatsunami: Wave of Destruction*—aired in 2000 as part of the BBC Horizon series. The programme, which featured both myself and Simon Day, presented the evidence for the initial sliding of the western flank of the volcano during its 1949 eruption; support for continued instability and movement of the flank provided by ground deformation monitoring; and a scenario for the likely scale and extent of the tsunami that would result from the eventual detachment of this part of the volcano and its rapid entry into the North Atlantic. Inevitably, while the science was sound, the treatment was somewhat populist, and the potentially devastating consequences hammed up. The result infuriated some of the tsunami and marine science communities and launched an often acrimonious debate that continues today. There have been further indications, however, of the potential seriousness of the situation at the Cumbre Vieja. In 2001, Steve Ward of the University of California, Santa Cruz, joined Simon Day in writing and publishing a paper that presented the first computer model forecasting the possible scale and extent of a tsunami arising from a future collapse of the Cumbre Vieja. For a worst-case scenario, involving the entry into the ocean of a mass of rock about comparable with Mauna Loa's Alika-2 slide, the model predicts that a huge initial dome of water, close to 1 km in height, would form within two minutes of failure, as the west flank pushed its way into the waters of the North Atlantic. After just 10 minutes, the model predicts a series of waves hundreds of metres high spreading out across a distance of 250 km, inundating the shores of the three westernmost of the Canary Islands; La Palma itself, El Hierro, and La Gomera. Over the course of the next hour, the coastal zones of the remaining islands—Tenerife, Gran Canaria, Fuerteventura, and Lanzarote—are swamped and waves as high as Big Ben encounter the north-west coast of Africa. Between three and six hours after collapse, the model envisages waves between five and seven metres in height coming ashore in Spain and on the south coast of the UK, while the coast of Brazil is on the receiving end of waves three times this height. At last, twelve hours after collapse, the model forecasts the deluging of the eastern Caribbean, while the flat, monotonous landscape of Florida faces a dozen or so waves up to 25 m high.

Whether or not this is a scenario that will play out in reality depends on a number of things, most critically the size of the collapsing mass, the speed with which it enters the ocean and the persistence of the tsunami at oceanic distances. It may well be that the landslide, when it occurs, will involve a smaller volume, which would result in a scaling down of both the resulting tsunami and the extent to which it remains destructive far from the source. The collapse itself is likely to be very rapid indeed; major volcano collapses that move slowly and benignly are unknown, either in the geological record or during historical times. The big argument among different elements of the tsunami community has raged around how rapidly the wave energy will be dissipated and how quickly the waves will shrink in size as they move outwards from their source. While Ward and Day envisage waves more than 20 m high striking the US coast, other tsunami modellers predict wave heights of just a few metres. Who is right we may never know—at least until the collapse happens. What is certain is that the level of death and destruction within the Canary Island archipelago itself, and probably across the entire north-east Atlantic, is likely to be unprecedented. However large the waves may be when they reach the Americas, it is a sobering thought that many thousands lost their lives in Thailand and Sri Lanka during the 2004 tsunami to waves that were only seven or eight metres high, or even less. Whether or not our changing weather and precipitation patterns as the world warms will encourage the Cumbre Vieja volcano to finally shed its western flank, or at least a substantial chunk of it, cannot be foretold. What is certain is that, as past experience tells us, where conditions become warmer and wetter, increasing instability and slope failure will follow. For many glaciated volcanoes too in Alaska, Kamchatka, the Cascades, the Andes, and elsewhere, which are already experiencing higher temperatures and melting ice, the chances of collapse will be significantly raised.

The big splash

Water is not fussy about the nature of the slopes it contrives to make unstable and persuades to fail, and it is as effective at triggering landslides in non-volcanic mountain ranges as it is at promoting lateral collapse at many of the world's volcanoes. As is the case for crumbling volcanoes, determining a direct link between past climate change and the incidence of major landslide formation in non-volcanic terrain is far from easy and primarily for the same reasons: an incomplete record and difficulties in accurately dating the events, or even managing to date them at all. Nevertheless, episodes of intense landslide activity have been identified during the transition from glacial to interglacial conditions that occurred during the latest Pleistocene and the Holocene, in places as far apart as the UK, the Pyrenees, and European Alps, the Carpathian Mountains, the peaks of the Canadian Rockies, the Andes, and New Zealand's Southern Alps. Many factors are likely to have played a role, each and every one involving water somewhere along the line. The exposure of expanses of bare rock and debris as glaciers and ice caps retreated, the thawing of mountain permafrost due to rising temperatures, and increased precipitation rates associated with changing weather patterns are all likely to have contributed towards reinvigorated instability development and slope failure during post-glacial times. In some areas, rapid uplift and an increase in earthquake activity due to ice unloading may also have had parts to play, through deforming already destabilized slopes or giving them a good shake. With so much more water around at cold-to-warm climate switches, various options are available whereby slope stability may be reduced and landslides triggered. A straightforward one in high mountain ranges simply involves rising air temperatures melting the ice that previously physically bound together a rock face or slope, perhaps helped by expansion of the rock itself as it warmed up. Additionally, masses of loose debris, particularly where exposed by retreating ice, can be saturated and mobilized through persistent heavy precipitation, while rainwater or snow and ice melt can act to destabilize rock masses, particularly where strongly fractured. Before an unstable rock mass can be persuaded to collapse, however, something else has to happen; a so-called 'failure surface' needs to form upon which the future land-slide will occur. This may happen rapidly, perhaps over periods as short as days or weeks, or—prior to the greatest of volcanic land-slides—take decades, centuries or even longer. While none of us were around to observe the development of such failure surfaces prior to the giant mass movements of the past, we can get some idea of how the process operates from analysis of a notorious landslide in north-ern Italy that shocked the world in the early 1960s.

Prior to a wet night in October 1963, Vajont was a name known to few who lived beyond the Piave Valley, high up in the mountains of Italy about 100 km north of Venice. The following day, Vajont was infamous, a name known around the world, a name synonymous with catastrophe and with incompetence. During the post-war period, when Italy—like many countries brought to their knees by the conflict—was struggling to build a better future, economically and socially, one of the key goals was to develop energy and water supplies in order to support industrial growth. With its mountainous terrain and high precipitation rates, the Alpine north had been looked to for some time as one of the obvious regions to improve energy and water provision, with plenty of gullies and gorges suitable for damming, and numerous valleys ripe for conversion to holding reservoirs.

A relatively inconsequential trickle, called the Vajont, seemed per-fect. The small watercourse flowed into the larger Piave River via a spectacular narrow defile, a good 300 m high, behind which the ter-rain opened out into a basin that would make a perfect reservoir. All that was needed was to block off the gorge with a dam containing turbines and hey presto—a simple and easy boost to the region's energy and water supply. So perfect was the set-up, in fact, that dam-ming of the Vajont had been proposed as early as the 1920s. The plans, however, had gathered dust until Mussolini was toppled in

1943, and even then it was another 14 years before construction of the dam began. In 1956, regional energy company SADE (Società Adriatica di Elettricità)—owned by former Finance Minister to the dictator, Count Giuseppe Volpi—began work on what was to become the world's tallest dam. Three years later, the narrow gorge through which the Vajont flowed into the Piave Valley was blocked by a towering 262 m high mass of concrete, as thick in places as three London buses laid end to end and three-quarters as tall as New York's Empire State Building.

Even before the reservoir behind the dam began to fill, however, the problems started, and during construction of a road along the southern margin of the planned reservoir, new cracks were found in the rock that hinted at growing unrest of Mount Toc, whose northern face formed the southern margin of the Vajont Valley. Serious concern about the stability of Mount Toc was expressed to SADE by a number of independent scientists, including geologist Eduardo Semenza—son of Carlo, the engineer who planned the dam. Notwithstanding these warnings, SADE ploughed ahead with the project and began filling the reservoir in February 1960. Even the collapse of close to a million cubic metres of rock into the reservoir later that year, together with a number of smaller landslides, led to only a temporary halt in proceedings. The response of Mount Toc was rapid and chilling. Within eight months, with the water 170 m deep, the north face of the mountain started to inch towards the reservoir, opening up a 2 km long M-shaped crack half a kilometre above the valley floor, which marked the detachment of a huge mass of rock from the rest of the mountain face. This too, failed to halt the project, driven ever onwards against any and all safety worries by political and economic pressures. Over the next three years, the crack continued to widen by a few millimetres a day. Crucially, when the reservoir levels were periodically

lowered the rate of crack opening slowed, leading to the false—
and ultimately tragic—assumption that the movement of the
northern face of Mount Toc could be tamed by regulating the depth
of water in the reservoir. Buoyed up by this misconception, the
push continued to fill the reservoir to full capacity and water levels
rose steadily into the autumn of 1963. By early September, with the
water level at 245 m, the rate of widening of the crack had increased
sharply to more than 3 cm a day, at last fostering concern among
the engineers responsible for the dam, who started to drain the
reservoir in an attempt to slow the accelerating movement—but it
was too late. This time there was no slow-down in the rate of opening, which, even with a 10 m drop in water levels, exceeded 20 cm a
day by 8 October.

The finale to this sorry story came at just after half past ten on
the wet evening of 9 October, when more than one-quarter of a
billion tonnes of rock detached itself from Mount Toc and, within
45 seconds, crashed into the reservoir; filling it entirely and hurtling more than 150 m up the opposing slope. With nowhere else
to go, the 115 million cubic metres of water in the reservoir—a
volume about equivalent to 72,000 Olympic-sized swimming pools—
was pushed over the dam as a wall of water—a splash-wave 150 m
high, plunging in seconds onto the town of Longarone that nestled
close to the foot of the dam and onto its unsuspecting citizens. The
weight of the water and the titanic force of the cushion of air it
pushed before it led to the annihilation of most of the town, along
with the neighbouring villages of Pirago, Villanova, Rivalta, and
Fae. Just a few minutes after Mount Toc's journey into the reservoir,
that part of the Piave below the dam was a boulder-strewn wasteland of mud scoured of every living thing, and 2500 men, women
and children were dead; the remains of many were never identified
or even found.

Fig 20. The Vajont Dam and 'before' and 'after' images of the northern Italian town of Longarone. The town and neighbouring villages were literally wiped from the face of the Earth by the giant flood precipitated by the collapse of part of Mount Toc into the reservoir behind the dam.

Fig 20. (*continued*)

Cracking up

How did the engineers and others involved in the Vajont Dam project get it so wrong? Forgetting, for a moment, the role of external pressures arising from the wishes of the construction company and the government to have the dam and reservoir up and working, the scene for a catastrophe was set once it became apparent that changing the water level in the reservoir could influence the rate of movement of the unstable mass. The engineers' interpretation of this relationship was that as water levels rose, so pore-water pressure in the rock making up Mount Toc increased, leading to accelerated movement and widening of the cracks. Conversely, when water levels were reduced, pore pressure in the rock fell, thus slowing the sliding of the unstable mass. Convinced that this was what was happening, the engineers felt that they could keep the movement of Mount Toc within safe limits by raising the level of the reservoir slowly and carefully. They were wrong on all counts.

It was not until 40 years later, however, that volcanologist and landslide expert Chris Kilburn, a colleague of mine at University College London, along with Dave Petley of the University of Durham, were able to shine a light on what was really going on. Far beneath Mount Toc's northern face, water from the filling reservoir was percolating into a clay layer around 200 m below the surface. At the pressures encountered at these sorts of depths, water is capable of corroding the rock, weakening the tips of existing cracks and encouraging their growth—a process known as 'slow cracking'. Given opportunity and time, these cracks will coalesce to form larger discontinuities, but as long as these are largely isolated from one another, deformation of the destabilized rock mass will take place slowly. As coalescence continues, however, and fractures start to join up, movement accelerates until ultimately a single discontinuity, or failure plane, is formed, completely detaching the deforming rock mass above from the stable rump below. At this point, the process is unstoppable and sliding occurs catastrophically. At Vajont, as reservoir levels were raised, so water entered the clay—which at the depths encountered was hard and brittle rather than soft and squidgy. The clay slowly expanded and opened up new cracks that progressively joined up. When water levels were reduced, cracking stopped or slowed—but the cracks didn't go away. Once levels were raised again, growth and coalescence of cracks continued, so altering the water levels did nothing to help increase the stability of the slope. All it did was limit the *rate* of the slow cracking process and temporarily postpone the inevitable. By the time the reservoir level had reached 245 m the fractures within the clay layer had joined up to a degree that failure of the slope was certain and the fate of the population of Longarone already sealed.

The slow cracking process is likely to have played a significant role in triggering large landslides during past episodes of abrupt climate change, most especially in the post-glacial period when ice-sourced meltwater, increased precipitation, and elevated levels of surface run-off were leading to the formation and expansion of many new bodies of water along the former margins of the ice sheets and in formerly glaciated mountain regions. Rising water levels would have provided the perfect conditions for destabilising adjacent slopes and rock faces and may also have promoted higher levels of instability around the margins of the depleted ocean basins. Overtopping and erosion of retaining topography due to landslide-triggered waves may even have caused or contributed towards the draining of some of the gargantuan pro-glacial lakes, such as Agassiz.

But to return to Vajont. Incredibly the dam was largely untouched by the huge wave that crashed over the top, and it stands today like a giant tombstone looking down on the rebuilt town of Longarone and

the long lines of victim's graves in its shadow. Perhaps even more surprisingly, the dam is fulfilling at least one of the roles for which it was
constructed, although admittedly to a degree far below that originally
envisaged: using water flowing through a by-pass tunnel bored in
1961, it is actually producing a small amount of electricity.

The Vajont tragedy teaches us lessons that we would do well to
remember as we ponder how to manage climate change. Human
interference in the natural world has consequences that are usually
surprising and often unpleasant, and 'taming' nature is nearly always
far harder than it first appears. In this context, it is worth observing
that at the global launch in 2008 of the International Year of Planet
Earth, UNESCO underlined the Vajont debacle as one of 'five caution-
ary tales' arising from the failings of engineers and geologists.

In the ocean something stirred

With water playing a key role in fostering the formation of giant
landslides on land, it would be a surprise if the wateriest of all
environments—the oceans—were immune to the destabilisation and
collapse of slopes. Indeed, some of the greatest landslides so far identified occurred in the submarine environment, their formation promoted by any one of a variety of triggers, or a combination thereof.
Once again, there is plenty of good, hard evidence that the frequency
of landsliding beneath the oceans is far from constant and is influenced by abrupt changes in our planet's climate. Given the huge
changes in global sea level associated with switches from glacial to
interglacial conditions and back, in combination—at high latitudes
at least—with all sorts of crustal disturbances along the ocean margins as the ice sheets ebb and flow, this should hardly be a revelation.
The best studied of all mammoth submarine landslides is a good
example and is well worth examining a little more closely.

Not much more than 8000 years ago, during the early Holocene, something stirred in the huge pile of glacial sediment dumped into the North Atlantic off the west coast of Norway. For one or other of a number of reasons addressed later, a vast chunk of sediment, with an approximate volume of 3500 cubic kilometres, detached itself from the submarine continental shelf, cascading downslope, spreading out and eventually coming to rest on the deep ocean floor after reaching more than halfway to Iceland. The sediment slide left behind a collapse scar close to 300 km wide—just a little less than the distance from London to Manchester or New York to Boston—and involved an area of sea floor one-third the size of the land area of the UK. A transfer of mass within the marine environment on such a spectacular scale could not be expected to happen without violently disturbing the ocean in the vicinity, and in this regard the so-called Storegga Slide did not disappoint. Distinctive sand layers preserved in the peat of the Shetland Islands—the UK's northernmost outpost—provide clear testimony to this. Discovered in the early 1990s by UK Quaternary scientist David Smith, and others, the sandy deposits are now recognized as having been emplaced by a tsunami triggered by the slide; the sand dredged up from offshore and dumped on top of the peaty landscape by waves that reached heights of up to 25 m as they flooded ashore. This is similar to the maximum run-up of the 2004 Indian Ocean tsunami, encountered in Sumatra close to the earthquake epicentre. Investigative studies since have found evidence for the Storegga tsunami at several locations along the Norwegian coast—where the tsunami run-up seems to have been between 10 and 12 m—and down the east coast of Scotland and northernmost England, where run-up heights were lower at 3–6 m. The reason that deposits have not been found elsewhere is either that no-one has looked closely enough or that conditions were not suitable for preservation of recognisable deposits. There is absolutely no doubt that the tsunami, which would have been massively damaging if it had happened today, would not have been limited to the vicinity of Norway and the northernmost UK, and must have inundated most coastlines in the North Atlantic region. It does seem that the closer you look the more you find, and more recent surveys of the coastal peatlands of the Shetlands have identified more sandy deposits that suggest that the islands were on the receiving end of two further, albeit smaller, tsunamis. One occurred around 5000 years ago and a second during Roman or early Medieval times; both were almost certainly sourced by further submarine slides somewhere in the North Atlantic.

The fact that three tsunami deposits are preserved in this small group of Scottish islands alone suggests that the Storegga Slide is far from unique, and this is indeed the case. The world's submarine environment is littered with the deposits of prodigious landslides, some sourced by island and coastal volcanoes, but most originating below sea level around the margins of the continents. Quite often, regions or territories that show evidence of particular instability can be recognized, where a number of major landslide events have occurred over time. The Norwegian continental shelf is a case in point, with three other major submarine landslides, in addition to Storegga, identified. Those parts of the submarine environment that reveal evidence for repeated landsliding have a number of things in common. Typically, there is a thick supply of sediment, providing a ready source of landslide material, and a sloping sea floor that helps to increase instability and that gives added impetus to a slide following its initiation. Submarine environments that are particularly prone to landsliding also tend to be pretty dynamic, perhaps characterized by an especially large input of sediment, as occurs at river deltas, or by high levels of earthquake activity, which is a feature of the deep ocean trenches that coincide with the world's subduction zones. Going back in time, the large rises and falls in global sea level and the subsidence and uplift of adjacent land masses, associated with ice sheet advance and retreat, would have provided an added dimension to the dynamism of landslide-prone marine environments.

Preparing for a fall

There seems little doubt that the Storegga Slide occurred as a direct consequence of climate change; the rapid passage from full ice age to the warmer world of the Holocene supplying the perfect circumstances. As addressed in the previous chapter, Norway, along with its Scandinavian neighbours and much of northern Europe, spent the last glacial period buried beneath a vast sheet of ice that, at its greatest extent, covered an area of close to seven million kilometres and achieved a thickness, in places, of more than 2.5 km. During this time, two things happened that would set the scene for the formation of the Storegga Slide once the ice had gone. First, the lithosphere beneath Norway was forced down under the weight of the ice by a good several hundred metres. Secondly, the currently submerged continental shelf was exposed due to reduced sea levels and covered by fast-moving glaciers that dumped ever-increasing piles of coarse, ice-pulverized, sediment and debris across the shelf and down onto the continental slope that pointed the way to the deep ocean floor. By 8000 years ago, the main body of the Scandinavian Ice Sheet had long gone, sea levels were almost at modern levels and the Norwegian continental shelf was once again submerged. But the lithosphere previously imprisoned beneath the ice was far from back to its normal interglacial level and continued to strive towards a new equilibrium now that the heavy carapace of ice had melted. As discussed previously, this entailed the reactivation of faults that had been quiet during the glacial period, and a resulting marked increase in earthquake activity. I think we can all make an educated guess about what happened next. The added load of a thick pile of glacial sediment to the clay-rich floor of the continental shelf promoted instability through weakening the underlying clays by raising pore-water pressures therein. All that was needed to open up a failure surface within the clays and set the slide going was an extra nudge, and this was provided by seismic shaking.

Earthquake activity, linked to the post-glacial uplift of Scandinavia is, then, the preferred mechanism whereby the Storegga Slide was triggered, and it seems to have happened more than once. Over the past half a million years as the ice sheets waxed and waned in response to changes in our planet's orbit and orientation in space, so the site of the most recent landslide was also the source of a number of other slides, their timing matching the glacial–interglacial cycles in a similar manner to the Holocene event. At high latitudes, the link—or at least one link—between climate change and submarine landslides seems now to be well established. During glacial periods, when sea levels are depressed, large volumes of sediment arising from the erosive power of glaciers are delivered to the ocean margins by ice sheets. During interglacials, when the ice sheets melt and contract, landslides in the sediment piles are triggered by shaking due to earthquakes linked to post-glacial uplift.

Norway is not the only place where giant submarine landslides promoted in this way are encountered. Others have been identified close to the contested UK rocky outpost of Rockall and off Ireland. Across the other side of the North Atlantic, 24 submarine landslides are recognized, adjacent to the location of the former North American Ice Sheet, most of which occurred during the last ice age and in the succeeding Holocene. Further south, off the Atlantic margin of the USA, more than 50 submarine landslides are encountered, most of which were probably similarly formed sometime during the last glacial period or in the Holocene. Seismic shaking caused by earth-quakes linked to the post-glacial rebound of the North American Ice Sheet is, once again, the favoured trigger for these slides. The Storegga model does not provide the whole story, however, and it can't be used to explain all the world's great submarine landslides. In particular, the situation is more complex at middle and low lati-tudes that were not glaciated, where a range of different factors can lead to destabilisation and sliding. Earthquakes are still likely to be an important landslide trigger, but related to local tectonic circum-stances rather then to rebounding lithosphere. Domes of the salty mineral known as halite, rising into and punching through conti-nental margin sediments due to its low density, may also have had a role to play in destabilising slopes and encouraging sliding. Other possible triggers include repeated loading caused by storm waves or even tides.

Still, as at higher latitudes, many submarine slides further from the poles seem to have occurred at the time of climate transition that characterized the end of the last ice age. In this context, rising sea levels appear to have had a major role to play through inundating and destabilising previously exposed continental shelves and maximising opportunities for slope failure. Even at times when the Earth remained in the long-term grip of bitter cold, rapid—though temporary—rises in sea level, associated with short-duration warm episodes, known as interstadials, seem to have been able to promote instability and fail-ure at ocean margins. A similar effect is apparent during post-glacial times, with a number of major slides off Mauritania in north-west Africa seemingly caused by climbing sea levels destabilising sand dune fields that had built out onto the continental shelf when sea lev-els were at their lowest. Looking at the broad picture of submarine landslide formation over time, it is clear that most occur at times when sea level is on the up or at least stable. What this might mean for the stability of the ocean margins in an anthropogenically warmed world, I will look at in the concluding chapter.

It's a gas

One element of submarine landslides that I have yet to address in detail is the possible influence of gas hydrates in their formation, which was touched upon in chapter 2. Gas hydrates are solid ice-like mixtures of water and gas—usually methane—locked away in marine sediments and imprisoned under Arctic permafrost in colossal quantities. Marine gas hydrates are the repositories of unimaginable quantities of carbon, amounting to two and a half times that already in our atmosphere. In the past they have been implicated in the onset of sudden and exceptional warm spikes in the Earth's climate, most notably at the Palaeocene-Eocene Thermal Maximum (PETM) a little less than 56 million years ago. Where pressures and temperatures are appropriate, marine gas hydrates will reside happily and indefinitely in solid form in the great piles of sediment that have accumulated over the aeons along the margins of the ocean basins. But they are time bombs. Modify the conditions of their environment, and they transform spontaneously from solid to gas, pumping their huge load of carbon-bearing methane into the atmosphere.

Gas hydrates are somewhat enigmatic beasts, some aspects of which remain poorly understood. The water and methane (and less commonly, carbon dioxide, hydrogen sulphide, and other compounds) that make them up are not tightly combined to form a chemical compound, but loosely hitched to one another in such a way as makes the transition from solid to gas, and vice versa, relatively easy to accomplish, simply by changing the ambient pressures or temperatures. For gas hydrates to form in the first place, a sufficient supply of water and gas needs to come together under conditions in which pressures are high enough, or temperatures low enough, or sometimes where the two meet to provide the perfect 'Goldilocks' environment. Under ice age conditions, when ocean temperatures were significantly lower than at present, it is reasonable to suppose that marine gas hydrates were able to form at shallower depths than they can today. It is also logical to reflect that such shallow deposits would have been particularly susceptible to the rapid warming of the oceans that characterized post-glacial times.

The methane that is the key component of most marine gas hydrates is formed by a number of processes operating in the depths of the seas, most notably through decomposition of organic debris, such as the mass of accumulated remains of dead plankton, and by the conversion of carbon dioxide contained in the sediment that floors the oceans. Because both temperature and pressure control the stability of gas hydrates, a sufficiently large change in either can promote a sudden switch from solid to gas. On a warming planet this provides us with an interesting situation. As the waters of the world's oceans heat up so the threat of marine gas hydrates becoming destabilized and dumping their load of methane into the atmosphere increases. On the other hand, as higher global temperatures melt increasing amounts of ice at high altitudes and latitudes, and sea levels climb as a consequence, the increased load will raise the pressure acting on the sediments containing gas hydrates. This would act to counterbalance the destabilising effect of rising sea temperatures. As it is highly unlikely that the effect on marine gas hydrates of warmer seas will be exactly matched by deeper seas, looking ahead it seems as if a race is on the cards to see which will win out. If the rate at which ocean temperatures rise is significantly greater than that at which they deepen then the odds favour the breakdown of gas hydrates and the release of quantities of methane to the atmosphere likely to ramp up global temperatures further and faster. Should rising sea levels take the lead, then the stability of marine gas hydrates could be assured, at least for a time.

One area of interest, which is relevant equally to past and future breakdown of marine gas hydrates, addresses exactly how the methane they contain is released. Does it just bubble upwards through the sediment and the ocean waters before eventually reaching the atmosphere, or is some other mechanism involved? One idea that has attracted a fair bit of attention in recent years incorporates a role for submarine landslides. A sufficiently large submarine landslide, so the theory goes, would be capable of triggering a breakdown of gas hydrates beneath by removing the overlying sediments, thereby instantly to all intents and purposes reducing the pressure acting upon them. Once it gets going, this is envisaged as a self-feeding, runaway process whereby the energetic release of methane gas acts to increase the area of slope failure, which in turn exposes further gas hydrates to lower pressures. Among others, the idea of a link between past marine gas hydrate breakdown and submarine landslides has been driven forward by Mark Maslin. A Quaternary scientist at University College London, Maslin and his collaborators point to a number of short-lived carbon isotope excursions (changes in the relative carbon isotope make-up) in marine sediments during Earth history, which they suggest owe their existence to the release of large volumes of the lighter carbon isotope, C^{12} which, as discussed in chapter 2, is regarded as a signature of carbon sourced from gas hydrate methane. One such excursion is coincident with the Chicxulub impact event, which defines the break between the Cretaceous Period and the Tertiary, 65 million years ago, and which brought about the demise of the dinosaurs and countless other life forms. According to Maslin, and the previously introduced Simon Day, such was the violence of this cataclysmic collision with a 10 km chunk of rock from deep space that

the associated planet-wide shaking triggered the widespread formation of submarine landslides. Across the world, the removal of large
masses of sediment in these landslides is charged by Maslin and Day
with causing sudden drops in the load pressure acting on gas hydrates
contained within, causing them to switch near-instantaneously from
solid to gas, releasing somewhere between 300 and 1300 billion
tonnes of methane into the atmosphere. While attractive, this idea is
also speculative. Nevertheless, it may account for the large rise in
atmospheric carbon dioxide levels that followed the impact event.

As addressed in chapter 2, there is strong evidence for a contribution from marine gas hydrates towards the sudden and severe warming event that characterized the PETM. In support of a submarine
landslide role in the destabilisation of gas hydrates at this time, the
Maslin school points to evidence of large-scale slope failures along
the western margin of the Atlantic Basin during the PETM. Recently,
it has been speculated that methane releases from marine gas hydrates
may also have played a role in the switchback climate of the Quaternary, which has seen our world shunted repeatedly from bitterly cold
to balmy during the last couple of million years. Once again, submarine landslides are presented as the key, the idea being that more of
these occur at times of rapid climate change, thereby leading to more
gas hydrate destabilisation and a dramatic rise in the amount of
methane in the atmosphere. This, in turn, would act to accelerate
warming further. As for the PETM and earlier periods in Earth history, however, the proposed link is liable to remain speculative for
quite a while. This is partly because there are still problems with
uncertainties in the record of atmospheric methane and partly
because accurately dating the ages of submarine landslides often
proves to be a particularly difficult challenge. Hard evidence is also
lacking in support of the converse idea, that marine gas hydrate
destabilisation at times when the climate is changing rapidly—either

due to falling sea levels or rising ocean temperatures—may actually act as a trigger for submarine landslides. It seems reasonable to link the formation of submarine landslides with the breakdown of gas hydrates contained in marine sediments in some way and there is plenty of circumstantial evidence that links the two. It may be some time, however, before a robust cause-and-effect connection becomes apparent. What is clear is that both the incidence of submarine landslides and the release of methane from gas hydrates are linked to periods in the past when the Earth's climate was being disrupted or in the process of rapid transition. Bearing in mind future projections for our world's climate in the coming century, we would be strongly advised to take note.

To recap: for a variety of reasons, a planet that is being transformed by climate change is also one in which slopes have a tendency to be less stable and more prone to failure. Simple water is once again the key, and just as it plays a role at such times in promoting earthquake activity and persuading volcanoes to erupt, so it is the ultimate driver of mass movements at the Earth's surface. A planet that is warm is also one where water is plentiful, where sea levels are rising; and where the ice that once held together the faces of mountains is going or gone. Beneath the waves, across the high mountain ranges and on the flanks of hundreds of active—and not so active—volcanoes, this is a world on the move.

During the climate megaflip that saw the late Pleistocene world of ice replaced—overnight in geological terms—by the mild conditions of the Holocene, melting ice, increased rainfall, filling ocean basins, and expanding glacial lakes acted together to dramatically increase the rate of erosion, mobilize slopes of loose debris, detach rock faces, raise pore-water pressures, and promote the growth of failure surfaces. This is a world that is now lost to us, but one that may well rear its head once again as the century progresses. While it is highly unlikely that anthropogenic climate change, however bad it gets, will elicit quite so vigorous a response from the Earth's more extreme topography, any jump in the level of slope failure and landslide formation could be both extremely damaging and widely lethal. Already, there are disturbing signs that rising temperatures are driving increased landslide and avalanche activity in mountainous terrain, especially at high latitudes. Perhaps this is the forerunner of far worse to come, but more on that subject in the final chapter.

6

Water, Water, Everywhere

The Earth is a water world, perfectly placed in relation to our Sun to ensure that this simple but magical compound can thrive here in all three of its forms. Swap places with Venus—a mere (in astronomical terms) 38 million kilometres closer to the solar furnace—and the baking temperatures would permit its existence solely in vapour form. Head in the opposite direction to Mars, and by far the dominant mode of this unique union of oxygen and hydrogen is the iron-hard ice that probably only survives in the polar regions. While there is plenty of evidence from surface features that water, in its liquid form, may once have been almost as common on Mars as it is on Earth, those days are long gone. Our own world stands out in stark contrast to both of our neighbours; a blue jewel that, when viewed from the depths of space, owes its stunning beauty to the cover of liquid water that obscures close to three-quarters of its topography. This is augmented by the sparkling sheets of ice that cap the top and bottom of this remarkable planet and the swirling white clouds of water vapour that drape the continents and the seas. But for all its ubiquity, water still makes up only a tiny fraction (about 0.023 per cent) of our world's total mass, which makes its importance to everything and everyone on the planet all the more extraordinary. Needless to say, water is a key prerequisite to the formation and sustenance of life—at least as we know it. Through the hydrological cycle it redistributes itself continuously between the atmosphere and the oceans, the land and its ubiquitous cover of greenery. By means of flood and drought, it can destroy the lives and livelihoods of millions, yet in a more benign guise it irrigates our crops and provides our vital waterways. While its erosive power can reduce jagged mountain ranges to flat plains, it is a vital lubricant of the tectonic forces that act to push them back up again. Water is something that, quite literally, we can't live without. As we have seen in previous chapters, it also plays a pivotal role in our world's ever-changing climate and in the manifold responses to it of the crust beneath our feet

The ups and downs of the Earth's oceans

Together, the weight of all the Earth's surface water adds up to an unimaginable number of tonnes—something like 1.35 followed by 16 zeros in fact. But where does it all come from and when did it appear? The source of the Earth's water remains a bit of a mystery. One theory holds that all the water needed to form the oceans was incorporated into the Earth as it accreted, in the form of water-containing minerals. As the Earth cooled, so water in vapour form 'outgassed' from the planet's interior through volcanic vents, condensing rapidly to form the earliest oceans. Fascinatingly, despite the enormous volume of the oceans the vast majority of our world's water is still locked away deep within its interior, where around 10 times—or perhaps as much as 50 times—the amount that occupies the ocean basins is stockpiled. An alternative model envisages the coming together of a bone-dry Earth, the water being added later, maybe through collisions with comets. The recent recognition, however, that sufficient water existed within the inner solar system at the time the Earth was sweeping up the debris it needed to build itself, means that our world could quite easily have incorporated enough water in the form of countless molecules of H_2O adhering to particles whose destinies were to become part of our planet.

Whatever its origin, 97 per cent of our planet's surface water is now incorporated within the immense repository that makes up the world's oceans. Together, this contains around 1.3 billion cubic kilometres of the stuff, just about sufficient to fill to the brim a colossal vase, around 1000 km wide and with sides equally high. Throughout Earth history the level of the world's oceans has been on something of a switchback ride. While we have only limited information about global sea levels during Pre-Cambrian times, the picture starts to become clearer following the start of the succeeding Cambrian Period, a mere 542 million years ago. From this time, when the fossil record shows that the seas teemed with all forms of amazing and imaginative invertebrate beasts, some broad trends in global sea level can be gleaned, bound up not only with a changing climate but also with the disposition of the Earth's tectonic plates during their sedate progress across the face of our planet over hundreds of millions of years. Perhaps surprisingly, given the fact that the oceans at present cover more than 70 per cent of our world, sea levels today are lower than they have been for most of the last half a billion years.

The term 'eustatic' was coined by the 19th century geologist Eduard Suess to describe changes in global sea level, and to differentiate them from local or regional variations that simply reflected the ups and downs of adjacent land masses in response to tectonic or isostatic forces. London-born but Czech- and Austrian-raised, Suess (1832–1914) was a giant of the Earth sciences, who transformed contemporary views of Earth history and, in particular, of the variability of the oceans throughout geological time. Although hindered somewhat by the fact that development of plate tectonic theory lay some way in the future, Suess made a number of remarkably astute observations and predictions, including postulating the existence of an ancient ocean that once separated Africa and Europe—now known as Tethys—and recognising, on the basis of fossil ferns, that the southern continents at some point in the distant past must have been grouped together forming a 'supercontinent' that Suess named Gondwanaland. He is even credited with helping to develop the concept of ecology, in which regard he was the first to use the term 'biosphere'. But to return to eustatic sea-level change; here Suess's biggest contribution was in recognising that evidence for rises and falls in the level of the world's oceans could be traced across the planet and connected up. This ultimately allowed a picture to be painted of marine transgressions, during which ocean waters encroached onto the world's land masses, and regressions, which involved the retreat of the seas and the exposure of more land. This view has been refined in the light of plate tectonic theory and the application of modern methods for logging the migrations of ancient shorelines, so that we can now look back with reasonable certainty at how global sea levels have been behaving over time.

The main determinants of global sea level are first, variations in the amount of water locked away as ice at higher latitudes and, second, the area and depth of the ocean basins. Global sea level typically changes by more than 100 m as the planet cycles between glacial and interglacial conditions. In addition, at times when the ocean basins are wide and deep, they can hold a great deal more water that during periods when—for one reason or another—they are more limited in area and shallower. At such times, ocean waters with nowhere else to go spill over to inundate vast tracts of low-lying topography adjacent to the margins of the basins. This last happened around 80 million years ago, during the Cretaceous Period, when warm shallow seas, infested with ferocious sharks, crocodiles, and marine reptiles, and enlivened by ammonites of every shape and size, invaded much of Europe, North America, and Africa.

How much water the ocean basins can hold at any one time is closely bound up with the disposition of the Earth's continents as they sport themselves across the surface of the planet on the backs of the Earth's tectonic plates, at times gluing themselves together to form a supercontinent, and at others scattering themselves widely across the globe. Throughout geological history, sea level has tended to be low when the continents are clustered together, and high when they are spread far and wide. Thus, sea levels were low around 300 million years ago, coincident with the emergence of the super-continent of Pangaea, and again 300 million years earlier than that, when the continents converged to form the less well-known super-continent of Pannotia. In contrast, during the Ordovician and Cretaceous periods, when the continents were dispersed, global sea levels were high.

The explanation for this relationship is quite straightforward. Following the break-up of a supercontinent, the young oceans separating the drifting continental fragments would have been floored by hot new lithosphere, formed at the mid-ocean ridges where the tectonic plates move apart. The high temperatures in the lithosphere meant that its density was relatively low, and this in turn meant that it 'rode high', resulting in shallow ocean basins with limited holding capacities and sea levels that were elevated. As the continents drifted ever further apart and the oceans separating them became wider and wider, so the lithosphere that floored them became colder and denser and therefore sank deeper. By the time plate movements had brought all the bits and pieces of the continents together again, the giant ocean basin surrounding the new agglomeration would have had an impressive capacity, hoarding the ocean waters easily and ensuring that global sea levels were low. A special situation seems to have existed when the continents were grouped at or around the poles, as they are at the moment and have been throughout the Quaternary ice ages. Whenever they converge at the top and bottom of the world, they provide a substrate for ice sheets, thus contributing towards the extraction of colossal quantities water from the oceans and reducing sea levels further.

The sea-level roller coaster has a distinctive M shape, with two peaks and three troughs. Following the break-up of Pannotia in the late Pre-Cambrian, sea levels rose to a high around 450 million years ago, during the Ordovician Period, when they touched an astonishing 200m or more higher than they are today. As the continents came together again to form Pangaea, so global sea levels fell dramatically. By 250 million years ago, in the Permian, they reached a low point that just about matched that of today's oceans. The disruption of Pangaea saw levels rise again to about 170 m above today's values, peaking in the late Cretaceous, around 80 million years ago. Since then—and notwithstanding the wild swings in sea level over the past couple of million years as glacial episodes alternated with interglacials—it has been pretty much downhill all the way.

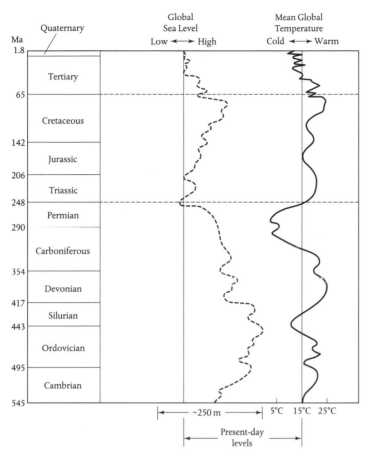

Fig 21. The sea-level roller coaster has a distinctive 'M' shape, with highs reached during the Ordovician and Cretaceous Periods and corresponding lows occurring in the late Pre-Cambrian, the Permian and the Tertiary/Quaternary.

Breakthrough!

In combination with tectonic movements and surface erosion, the wild swings in ocean levels have conspired to produce some of the most cataclysmic geomorphological events in Earth history. Land bridges have been exposed only to be subsequently and rapidly obliterated, while deluges and cataracts of unimaginable proportions have been created as rising seas have broken through or spilled over into adjacent regions of low topography. Probably the most incredible event took place a little over 5.3 million years ago, towards the end of the Miocene Epoch. Around a quarter of a million years earlier, the Strait of Gibraltar, through which Atlantic waters today enter the Mediterranean Sea, was finally—after a few transitory closures—sealed by tectonic forces. Isolated from replenishing waters and subject to a dry hot climate, the Mediterranean soon began to shrink through evaporation and eventually became completely desiccated. Putting today's Dead Sea and Death Valley in the shade, the Mediterranean was ultimately reduced to a vast, empty hole, more than 2.5 km below sea level, floored by a few desultory lakes of super-salty water. Everything changed, however, when the waters of the Atlantic reclaimed this desert wasteland, in a truly spectacular manner.

As a consequence of tectonic subsidence, perhaps aided by rising sea level and erosion, the Atlantic finally burst through in an event known as the Zanclean Flood, slicing channels 250 m deep through which the waters flooded into the empty basin. According to Spanish geologist, Daniel Garcia-Castellanos, and his co-workers, the Atlantic flooded through the gap at a rate one thousand times greater than the flow of the Amazon River. So rapid was the transfer of water, in fact, that the fledgling Mediterranean Sea may have risen by as much as 10 m a day, the entire basin being largely filled in a period as short as a few to 24 months. Tempting as it would be to imagine a giant Niagara or Victoria Falls taking centre stage within this singular geological drama, as is indeed sometimes represented, this would be far from the mark. According to Garcia-Castellanos and his team, it looks as if the replenishing waters would not have formed a mega-cascade, but would instead have poured into the Mediterranean down a low-angle ramp with a slope of just a few degrees. However, this in no way diminishes the stature of this extraordinary incident, and for many geologists it must surely be close to the top of the list of 'things to see if I had access to a time machine'.

A similar catastrophic deluge has been invoked for the flooding of the Black Sea just 7500 years ago, this time by seawater originating in the Mediterranean itself. The idea, first championed by geologists William Ryan and Walter Pitman in their 1997 book *Noah's Flood*, caught the public imagination primarily as a consequence of the authors' suggestion that the deluge may have been the trigger for the myth of the Great Flood, which crops up in a number of ancient texts including, of course, the Christian Bible and the Akkadian *Epic of Gilgamesh*. Recent research suggests, however, that the filling of the Black Sea may not have been quite so catastrophic after all and may have been a less spectacular and far more laid-back event than the flooding of the Mediterranean basin.

Although not on a par with rising Mediterranean levels following the Gibraltar Strait breakthrough, rapidly rising sea levels at the end of the last glaciation resulted in the eventual overwhelming of all or most of the land exposed when sea levels were depressed. One such area was the 1500 km wide land bridge known as Beringia, which joined Alaska to Siberia, and which allowed the traffic of large mammals between the two regions and, 12,000 years ago, the migration of Asian peoples from eastern Siberia to North America—the ancestors of the Native Americans. The inexorably rising post-glacial sea levels claimed other land bridge victims too, including the pre-Eurostar English Channel link that joined England to continental Europe and those that connected Australia to New Guinea in the north and Tasmania in the south. Higher sea levels also created islands out of Sumatra and Java (in modern-day Indonesia) and nearby Borneo, all formerly attached to mainland Asia. Looking ahead, there are no places under immediate threat of a breakthrough by oceans that, after a few thousand years of relative stability, are once again on the rise, this time due to climate-change-driven thermal expansion of seawater and, increasingly, by a contribution from melting glaciers and polar ice sheets. However, when and if a continued 'business as usual' emissions path commits us to losing the Greenland and West Antarctic ice sheets, and therefore to an eventual global sea level rise of more than 10m, the scene will surely be set for another serious assault on the world's shorelines, one that would almost certainly lead to some catastrophic incursions. The Netherlands beware!

Our planet throws its weight around

For many of us, the sea presents little more of a threat than a touch of sea-sickness on a choppy boat trip or a thorough soaking by storm waves on a winter walk along the promenade. This is far from the case, of course, for the hundreds of millions of people living on coastlines periodically battered by hurricanes or typhoons, or subject to large, tsunami-spawning earthquakes. For many of these people, the sea does provide a livelihood, but one that is frequently fragile and which can be erased in an instant. We don't have to look far to find examples of how the life-giving oceans have turned to crush those who are dependent upon them: Indonesia, Thailand, and Sri Lanka in 2004, New Orleans in 2005, and Japan in 2011. Tsunamis and storm-surges driven by powerful tropical cyclones are the ocean's shock troops, taking thousands of lives a year and periodically wiping out hundreds of thousands in single cataclysms. While developing countries are consistently worst affected, economic development and social sophistication does not provide sufficient immunity. Communities in the industrialized nations continue to pay a price for their proximity to the sea; evidence the more than 2000 lives lost in the Netherlands and the UK to a North Sea storm surge in 1953; the 1800 deaths incurred during the battering of New Orleans by Hurricane Katrina in 2005; and the 20,000 souls lost in the 2011 Japanese tsu-

Fig 22. The aftermath of Hurricane Katrina, which took 1800 lives when it struck New Orleans in 2005, provides testimony to the risk involved in living close to the ocean.

nami. With sea levels expected to rise throughout the course of the 21st century and far beyond, and with populations in coastal zones rocketing, this is a picture that can only get worse. The sea also exacts a toll far out from the shorelines, with towering rogue waves that can be higher than 10-storey buildings, capable of capsizing and sinking even the biggest ocean-going ships. Just how much shipping succumbs to rogue waves far from land is unknown, but year on year dozens of cargo ships vanish without trace on the high seas.

In addition to such common-or-garden hazards, we have seen in previous chapters how the changing gross characteristics of the world's oceans at times of rapid climate transition drive a completely different portfolio of hazards. This time it is linked to the solid Earth beneath rather than to a combination of the sea and the atmosphere above, and including everything from the destabilization of gas hydrates to submarine landslides, and from tsunamis to erupting volcanoes. Before addressing further the many and varied direct effects of changes in ocean levels and temperatures on potentially hazardous geological systems, it is worth stepping back and considering the broader picture, particularly in relation to times when our world was in the throes of the switch from glacial to interglacial conditions and back. These transitions saw extraordinary amounts of mass being shifted around the planet over periods as short as 10,000 years or less—a mere blink of an eye in geological terms. At the present time, the oceans hold more than 45 times the volume of water held in the polar ice sheets and other smaller areas of land ice. At the heights of glacial periods, however, an enormous volume of water was decanted from the ocean basins and used to build the great continental ice sheets and to expand smaller mountain ice caps and glaciers. At the Last Glacial Maximum this amounted to an astonishing 52 million cubic kilometres of ice or about 4 per cent of the total ocean volume. This is a small fraction, but nevertheless one that represents a staggering weight to be hauled around the planet over a very short period of geological time. Rearranging the Earth's mass on this sort of scale is even sufficient to cause adjustments in its spin. The magnitude 9.0 Japan earthquake of 2011, for example, shifted the Pacific Plate in the area to the west by 30–40 m; a transfer of mass great enough to cause our world to spin faster and to shorten the day by a little under two microseconds. Other spin adjustments in response to mass movement associated with great earthquakes were also detected after the

2004 Sumatra and 2010 Chile earthquakes. Major earthquakes can also alter the position of the Earth's figure axis—the axis around which the mass of our planet is balanced and which is separated from its N–S axis by about 10 m. The Japan event seems to have displaced this by about 17 cm, contributing towards a slight change in the way the Earth wobbles in its passage around the Sun.

The fact that major geological events can affect the Earth's spin, the length of its day, and the manner in which it wobbles in space, should not be unexpected. As observed in chapter 3, geodetic measurements that monitor the shape of the Earth extremely precisely show that its shape changes with the seasons as the atmosphere and the oceans constantly redistribute mass around the planet. This, in turn, is reflected in changes in the length of the Earth's day by a whole millisecond—an effect 500 times greater than that arising from mass changes associated with the Japan quake. In chapter 5, I described how reallocations of the planet's mass over the course of a year, arising from annual movements of surface water, have been linked to a seasonal response of the world's volcanoes, such that they are 'forced' to erupt by tiny changes in stress and strain within or beneath them. Given that today's population of active volcanoes seems to be responding to the broad-scale deformation of the planet associated with the movement of a water mass more than 5000 times smaller than that swapped back and forth between the oceans and ice sheets during major climate transitions, it would not be unreasonable to envisage a similar, if not greater, sort of response from the solid Earth at such times. Rather than driving an annual migrating pulse of deformation around the planet, the growth and decay of ice sheets involves a much slower change in the shape and spin characteristics of the planet. As ice sheets expand at high latitudes and more mass is concentrated closer to the spin axis, so the Earth's spin is progressively speeded up—in exactly the same way as a figure skater brings her arms closer to the body in order to rotate faster. Conversely, as the ice melts during transitions to interglacial conditions, and mass is transferred to the oceans further from the spin axis, so the speed at which the Earth spins is reduced. This equates to a gyrating figure skater stretching out his arms in order to rotate more slowly.

Such accelerations and decelerations in the Earth's spin rate affect the behaviour of all fluid systems, including the atmosphere, the hydrosphere, and—deep within the Earth—the asthenosphere where magma is generated. They also influence the deformation of the crust. In China, for example, scientists monitoring active faults in the hope of detecting earthquake precursors have noticed that tiny variations in the length of day, caused by spin-rate fluctuations over the course of decades, are linked to stress changes in the crust that result in several millimetres of movement on the said faults. The length of day variations charged by the Chinese team with promoting measureable movements along active faults add up to less than a couple of milliseconds. Compare this, then, with the four-second shortening of the day that accompanies the growth of the ice sheets, and the comparable lengthening attendant on their melting and the transfer of mass to the ocean basins. While occurring over far longer time periods, it would seem likely that changes in the state of stress and strain within the Earth's crust, associated with these much greater variations in length of day, had some influence on geological systems such as active volcanoes, earthquake faults and, perhaps, the movement of large, unstable rock volumes.

As well as affecting the speed at which our world turns on its axis, and therefore the length of day, the repeated transfer of mass between the ocean basins and the ice sheets, and back again, also resulted in modifications to the shape of the *Geoid*. This can be thought of as the invisible (or at least not detectable to the naked eye) form of the Earth defined not by its topography, but by its gravity. The Geoid is, to all intents and purposes, coincident with mean sea level and can be thought of as the shape of our world were it entirely covered in water. The Geoid, however, is not an absolutely perfect sphere. Due to centrifugal force linked to the fact that the Earth is spinning, it takes the form of an oblate spheroid—that is a sphere that has been flatted at the top and bottom and which bulges at its middle. This means that the Earth's diameter measured at the equator is a little over 40 km longer than that measured from pole to pole. In addition to this, the Geoid has a topography arising from differences in the pull of gravity at various points on the Earth's surface, which are in turn reflections of the distribution of mass within our planet and geophysical processes operating deep within its interior. If artificially exaggerated, the topography of the Geoid would make the Earth look more like a lumpy potato than a sphere, but in actual fact it is far more subdued than that described by the high mountains and deep ocean trenches, which define a maximum height difference of nearly 20 km. In comparison, the peaks and troughs of the Geoid vary in elevation by less than a couple of hundred metres across the entire planet. Not unexpectedly, the huge weight of the continental ice sheets that developed during the last glaciation had a significant impact on the shape of the Geoid, resulting in its depression beneath Europe and North America. This negative anomaly can still be seen today in the north-eastern part of North America, but is gradually disappearing as the once-imprisoned lithosphere continues to bounce back. At the same time, the plastic part of the mantle known as the asthenosphere is still flowing back to higher northern latitudes, having been displaced southwards by the weight of the ice sheets forcing the overlying lithosphere downwards at the last glacial maximum. This migration of mass associated with post-glacial rebound, in its own right, is estimated to still be having an effect on our planet's rotation rate, tending towards slowing it down by a little over half a millisecond a century. Taken together, stress and strain variations associated with the distortion of the Earth's form, alongside excursions of the asthenosphere and related adjustments in the Earth's rate of spin and length of day, can reasonably be expected to have had some influence on geological systems such as active volcanoes and earthquake faults at major climate switches.

Keeping a low profile

To say that quantifying the driving potential on latent hazardous geological phenomena of stress and strain changes within our planet, promoted by the cyclic redistribution of 52 million cubic kilometres of water, is not easy, would be an understatement. Probably the best we can surmise is that such a dynamic situation would seem to be conducive to a more lively response from poised and primed geological systems. A more straightforward way of pinpointing a potentially hazardous response of the crust to the wholesale reorganisation of planetary water is to focus on the behaviour of sea level and to look for ways in which changing sea levels can be shown to be implicated in modulating or triggering bursts of volcanic activity, encouraging faults to move, promoting the sliding of large piles of submarine sediment, or influencing the destabilisation of marine gas hydrates. Such effects have been examined, to varying degrees, in previous chapters. Here, however, I want to look more systematically at how different sea level states have been linked to geological activity that today would present a threat to human society and activities.

Ignoring the fact, for the moment, that the Earth was 6°C cooler, and that a significant fraction of the land area of the northern hemisphere was encased in ice, the biggest differences between early 21st century Earth and the world of 20,000 years ago, arise from the consequences of a 130 m difference in sea level. As mentioned previously, land bridges were exposed connecting parts of the world that have been separated by sea for at least the past 10,000 years. The shallow coastal shelves were also revealed, forming almost pancake-flat terrain averaging around 60 km wide. During so-called low sea-level stands, the world's coastal shelves acted as sinks for huge volumes of coarse sediment, either washed down by rivers or, at high latitudes, transported by glaciers. The accumulation of such piles of erosional detritus, dumped on top of finer, clayey, sediment amassed when the shelves resided beneath the waves during the last interglacial, set the scene for future slope instability—the sticky, mud layers, providing ideal failure and slip surfaces for submarine landslides such as Storegga, once sea levels had risen again.

This is not to say that at times of low sea level, when the land encroached in places up to 1500 km or more into Neptune's domain, the ocean margins were completely stable. Evidence exists for some very large submarine landslides close to the height of the Last Glacial Maximum, most particularly a huge Mediterranean slide identified and described in the late 1990s by a team from the UK's National Oceanography Centre in Southampton. The evidence for this slide takes the form of a 'turbidite'—a thick pile of sands and silts representing the final product of the passage into the abyssal depths of a large mass of sediment disturbed and detached from the continental shelf. So impressive is this particular turbidite that the prefix 'mega' has been attached, to highlight the fact that its dimensions, and therefore those of its source are exceptional. The megaturbidite in question is spread across much of the Balearic Basin, the area of deep sea that separates Spain and France from North Africa and which is bounded to the east by the islands of Corsica and Sardinia. The source of submarine landslide that fed the turbidite, which has a volume of around 500 cubic kilometres, lay somewhere to the north and seems to have been the biggest event of its type in this part of the Mediterranean Sea for at least the last hundred millennia. The most interesting thing about it, however, is its age, which, at 22,000 years, puts its formation close to the time of the Last Glacial Maximum, when global sea levels were at or near their nadir. As discussed in the previous chapter, this is somewhat out of kilter with many other large submarine landslides, which appear to favour rising sea levels. Neither is the Mediterranean megaturbidite unique, with other colossal submarine slides too linked to low sea levels. The enormous fan of sediment that encircles the great delta of the Amazon River, for example, hosts a number of gigantic submarine slides, each having a volume of up to 2500 cubic kilometres. The dating of these collapses is not very reliable, but some at least look as if they were formed deep within the last glaciation when sea levels were either low or falling. The Cape Fear and Currituck sediment collapses, off the Atlantic coast of the USA, also seem to have been triggered when sea levels were considerably lower than they are today.

Broadly speaking, earthquakes related to the post-glacial uplift of formerly ice-swamped land masses are regarded as constituting the primary trigger of major submarine sediment slides at times when the climate is warming and sea levels rising. Explaining such events at times when the planet is icebound, the level of the oceans dramatically reduced and the coastal shelves exposed is less straightforward. Of course, earthquakes will still happen on occasion and may represent a valid trigger. Another possibility is that the lowered sea level reduced the pressure on gas hydrate deposits trapped in sea-floor sediments, such that they underwent sudden conversion from solid to gas. This huge volume change could have exerted sufficient pressure to set off the sediment slides. In fact, just this mechanism has been proposed as a possible driver of the Mediterranean megaturbidite, and has been judged a prime candidate as far as triggering at least some of the giant Amazon slides is concerned. In the late 1990s, the huge sediment collapses off the Amazon Delta formed a focus of study for Mark Maslin, lauded US marine geoscientist Bilal Haq, and others. The landslide deposits here are spread across an area of more than 15,000 square kilometres, not far off the size of the Caribbean island of Jamaica; they comprise 50 trillion tonnes of sediment and are up to 200 m thick—about the height of 40 London double-decker buses stacked on top of one another. Of the four slides identified along the submarine continental slope off the Amazon delta, two were formed in the depths of the last ice age when sea levels were falling fast, one about 35,000 years ago and the other 7000–10,000 years later. Global sea levels at the time may well have been plummeting at rates as high as 15–25 m every millennium, as water was sucked from the oceans to feed growing ice sheets; this is quickly enough—according to Maslin and his team—to promote the breakdown of gas hydrate reservoirs within the marine sediments. As mentioned in earlier chapters, gas hydrates can be encouraged to switch from solid to gas by higher temperatures. Even if the temperature remains constant, however, dissociation can be accomplished by reducing the ambient pressure. A rapid and dramatic fall in sea level would be a perfect way of accomplishing the latter. Gas hydrates have been identified in the area today and environmental conditions would have supported their existence in the past. Certainly, Maslin and his colleagues are convinced that gas hydrate breakdown triggered by falling sea levels resulted in the mass failure of this part of the Brazilian continental slope. A similar mechanism has been proposed for submarine landslides in the Gulf of Mexico and off the Atlantic coast of the USA.

Eruptions and earthquakes in a world of ice

In addition to destabilising marine gas hydrates and so promoting submarine landslides, the low and falling sea levels associated with ice-world conditions would also have had an impact on the world's thousands of Quaternary volcanoes, the vast majority of which form islands, are coastally located, or sit within a few hundred kilometres of a shoreline. It seems to be possible for both rising and falling sea levels to promote eruptions at active volcanoes, the dynamism of a changing environment being the critical factor. On the basis of the behaviour of volcanic systems today, Steve McNutt's work on the timing of eruptions at Alaska's Pavlof volcano provides strong support for rising sea level being the trigger. On the other hand, Ben Mason and colleagues, in their recognition of a 'volcano season', observe that peaks in the eruption rate over the course of a year seem to coincide more often with falling sea levels than times when sea levels are on the up. Undoubtedly, there are a number of ways in which one can imagine that a primed coastal or island volcano can be persuaded to erupt by reducing the level of the adjacent or surrounding ocean. These include making the opening of magma paths to the surface easier to accomplish as a consequence of the reduced pressure acting on the edifice; the debuttressing of the seaward flanks due to removal of water mass, increasing the chances of collapse which, in turn, can trigger an eruption; and, probably most importantly, the likely reduction in pressure exerted on the magma reservoir within or beneath the volcano. For an island volcano, a 100 m drawdown in sea level results in a fall of 1 megapascal in the compressive pressure acting on the volcano. This is the equivalent of 10 atmospheres and represents a pressure change more than 500 times greater than that which seems to be responsible for modulating eruptions of Pavlof due to seasonal rises in local sea level.

One volcano that seems to have responded to falls in sea level during the Quaternary forms the Mediterranean island of Pantelleria. This small island, which sits between Sicily and North Africa, is still logged as active, and while its last eruption was way back in 1891, it will certainly erupt again at some point. The 1891 eruption was a submarine event of which no trace remains above the surface. Interestingly and amusingly, a similar submarine eruption exactly 60 years earlier and 50 km to the south resulted in the formation of an island—known as Ferdinandea—and launched a vigorous burst of sabre-rattling from the great world powers of the time. Hardly had the vent stopped spewing lava and ash before the governments of the UK, France, Spain and, of course, Italy, were making claims of ownership in respect of the tiny speck of cooling volcanic rock. The French named the island Julia, and the British provided the appellation, Graham Land, after First Lord of the Admiralty, Sir James Graham. In the following months, while warships were dispatched and governments blustered and bridled, the island slowly but surely succumbed to the erosive effects of the Mediterranean waters, and by the end of 1831 it had vanished beneath the waves. With nothing left to lay claim to, the jingoism rapidly evaporated, but with the island today lurking just six metres beneath the surface, a new eruption could see both the island and the sovereignty argument rejuvenated.

Meanwhile, a little to the north, the island of Pantelleria remains firmly in Italian hands and has shown little evidence of life recently, with just the single, late 19th century eruption testifying to its latent threat. During prehistoric times, however, the situation was very different, with eruptions big enough and violent enough to form two calderas, one a good 6 km across. What is interesting about these calderas—excavated 114,000 and 45,000 years ago—is the way in which eruptions that occurred after their formation coincide with low sea-level stands during the last glaciation. The margins of calderas are always weak zones, because they define fractures along which a central block of crust subsided following eruption and evacuation of the magma reservoir underneath. It is common, therefore, for later eruptions to be fed by magma squeezing up these fractures, and in this respect Pantelleria is no exception. The periods during which magma made serious use of these lines of weakness, however, does not seem to be random, coinciding with times when sea levels were at or near rock bottom: 67,000 years ago and again about 20,000 years ago, close to the height of the Last Glacial Maximum. According to Peter Wallmann and other members of the Stanford University team that worked on the volcano in the late 1980s, this should not be surprising, and the likely mechanism is actually quite straightforward. At times of high sea level, the greater load exerted by the Mediterranean waters surrounding the island compresses the marginal fractures of the calderas, making it more difficult for new magma to force its way up them to feed eruptions. As ice sheets grow and sea levels fall, so the pressure on the fractures gradually reduces until sea levels reach their lowest point. Under such conditions, any available magma is able to force its way upwards more easily and in larger volumes than when sea levels are higher. In this way, the growth and decay of ice sheets, via the rise and fall of the Mediterranean, looks to have played a key role in controlling eruptions on this island volcano.

It would be reasonable to assume that magma at other volcanoes around the world would also have made the most of lower sea levels and reduced pressures in order to push towards the surface and feed eruptions. Island volcanoes are best suited to this, but sea level lows would also have had a big influence on the activity of submarine volcanoes, notably those that populate the network of mid ocean ridges that crisscrosses the world's ocean basins. This vast array is made up of long, linear, zones of undersea mountains that stretch for a total of 65,000 kilometres—a distance almost one third of that separating our planet from its Moon—along which the Earth's tectonic plates are slowly but steadily moving apart. They are host to literally hundreds of thousands of volcanoes that never see the light of day, outnumbering by far those that occupy land. It does not take a particularly large leap of intellect to come to the conclusion that low sea levels should make the eruption of magma at submarine volcanoes easier. First of all, this reduces not only the load pressure exerted on individual volcanoes and their magma reservoirs, but also the weight pressing down on the asthenosphere beneath, where magmas have their ultimate home. As the asthenosphere rises upwards in response, so it starts to melt and generate new supplies of magma, capable of feeding more to the volcanoes above. The corollary is that low sea levels set the scene for mid ocean ridge volcanoes to squeeze out the magma they contain more easily *and* for the production of more magma to replace that lost during eruption. Actually measuring this effect is probably not possible, if only because our knowledge of the timing of submarine eruptions is poor to non-existent. Geophysical modelling suggests, however, that volcanic activity on the mid ocean ridge system may have been around one fifth higher at the height of the last ice age than it is today.

If reduced loading due to big falls in global sea levels can promote prodigious quantities of sediment to slide, gas hydrate deposits to break down and volcanoes to erupt, could undersea earthquake faults also be susceptible? As related in chapter 4, there are both natural and man-made examples of reduced water levels, respectively in glacial lakes and artificial reservoirs, promoting an increase in local earthquake activity. Might globally reduced sea levels have driven, in the same way, a planet-wide seismic response at the height of the last ice age? As might be surmised, building up a picture of the past movement of submarine faults is not an easy task, and it is simply not possible to construct a nice neat record of undersea earthquakes showing how this has changed between the Last Glacial Maximum and the present day. Once again we have to look to modelling to provide us with at least a part of the answer. Reducing the level of water above active plate boundaries by 100 m or more can be shown to result in a significant reduction of load pressure such that offshore faults would be expected to rupture more easily. The effect of this would likely be that faults that would have ruptured in the fullness of time, spread over an extended time frame, would be swayed to do so within a smaller time window, resulting in a clustering of seismic activity. Whether or not this resulted in a burst of major subduction-zone earthquakes, such as those that have struck recently in Sumatra, Chile, and Japan, we may never know—but see later in the next section for some of the most recent thoughts on this idea.

Upwards and onwards

Much as the switch to smaller shallower oceans that accompanies ice sheet growth clearly influenced geological activity, the degree to which it does so proves to be minor in comparison to the response of the geosphere in the post-glacial world of submerged continental shelves and brimming ocean basins. Following the 2011 Japan earthquake and tsunami, the internet once again sizzled with speculation that anthropogenic climate change was somehow implicated. As far as I am aware, however, no mechanism exists by which our changing climate, proceeding at its current rate, could trigger a giant subduction zone earthquake, such as that which spawned the Japanese tsunami. This was clearly a quake whose time—following a wait of more than a thousand years—had come, and which was driven by tectonic forces alone. Any suggestion that contemporary sea level rise—currently running at about 3 mm a year—contributed towards this event and the recent cluster of giant plate boundary earthquakes that have rocked Indonesia, Chile, and Japan, seems unlikely in the extreme. But what about during the late Pleistocene and early Holocene, when sea levels were going up, on occasion, more than 10

Fig 23. Despite much speculation, there is nothing to suggest that anthropogenic climate change was implicated in the 2011 Japanese earthquake and tsunami.

times faster than they are today? Might this have influenced the activity of submarine plate boundary faults, including those marking subduction zones?

A recent study by Karen Luttrell and David Sandwell of California's Scripps Institute of Oceanography sheds some light on this issue and comes up with some interesting conclusions. The Scripps pair examined the impact of a 120 m sea level rise, simulating the post-glacial situation, on two types of plate-boundary fault: first, an off-shore subduction zone fault comparable to that which generated the great quakes of the 2004 to 2011 period, and second, an onshore trans-form fault. Unlike at a subduction zone boundary fault, which marks the join between an overriding plate and a subducting plate that is thrusting down beneath it, a transform fault forms the junction of two plates that are scraping sideways past one another, the classic example being California's San Andreas Fault. The relative movement of such pairs of plates is never smooth, and is typically characterized by long periods when the intervening fault is 'locked' and strain is building up, punctuated every now and then by earthquakes that release the strain of decades or centuries almost instantaneously.

Key to how and why a large rise in sea level might affect plate boundary faults, both offshore and onshore, is the manner in which ocean loading affects the broad area where pairs of plates meet. Not unexpectedly, the added load of an extra 100 m or more of water causes the lithosphere to bend along the join between the land and the ocean, which in turn modifies the stresses influencing land-based faults. In the simplest terms, the consequences are that times of high sea level promote more earthquakes at faults on land at the expense of those occurring on faults beneath the sea, the reverse being the case at low sea level stands. According to Luttrell and Sandwell, for example, the bending of the lithosphere due to a large rise in sea level would act to 'unclamp' California's San Andreas Fault, notorious for nearly obliterating the city of San Francisco in 1906, thereby promot-ing its rupture. They come to a similar conclusion in relation to the North Anatolian Fault, which slashes across northern Turkey and was responsible for more than 17,000 lost lives in the 1999 Izmit earthquake. Harking back to the filling of the Black Sea basin addressed earlier in this chapter, the Scripps researchers observe that this was a time when rapid ocean loading stood a fair chance of causing the entire fault—which runs close to the southern margin of the basin—to 'unzip', launching an earthquake 'storm' that would have migrated along the length of the fault.

At subduction zone faults located in deep water offshore, the large post-glacial hike in sea level would be expected to reduce the likelihood of their rupturing within any given time frame. In some circumstances, however, it looks as if the situation is more complicated. According to Luttrell and Sandwell, subduction zone faults whose active parts extend in a shallowly dipping plane beneath the overriding plate are more likely, when sea levels are high, to rupture at greater depths and therefore further towards the interior of the overriding plate. In contrast, when sea levels are low, fault rupture and earthquake generation occurs nearer the surface but offshore. These results have important implications for one of the world's best known and, potentially, most dangerous subduction zone faults, which parallels much of the west coast of North America.

Residents of coastal communities all the way down from the Canadian province of British Columbia to northern California, must have watched the terrifying images of destruction and desolation in northeast Japan, following the 2011 quake, with particular trepidation. Just over 300 years ago, the same scenario of geological mayhem was being enacted on their side of the Pacific. In 1700, the Cascadia subduction zone (CSZ), a 1000 km long fault marking the junction between the Juan de Fuca tectonic plate and the much larger North American plate beneath which it plunges, was also the source of a giant megathrust earthquake. The fault unzipped along its entire length, triggering a massive quake registering a magnitude of between 8.7 and 9.2. Huge tsunamis battered much of the west coast, and—in a mirror image of the 2011 event—were persistent enough to cause damage in Japan on the other side of the Pacific.

Waiting for the Big One

Few people were around to observe the extreme power of the 1700 quake, and reliable contemporary reports are almost non-existent. Detailed geological studies undertaken by the likes of Brian Atwater of the University of Washington, author of *The Orphan Tsunami of 1700*, and Chris Goldfinger of Oregon State University reveal, however, that this was just one of many great CSZ quakes that shook the region during prehistoric times—an astonishing 41 earthquakes in excess of magnitude 8.2 in the past 10,000 years in fact. This translates to one every 500 years or so. You might expect this information to bring sighs of relief from those living close to the Pacific coast of North America only 300 years after the last Big One, barring two things: first, the Earth never operates like clockwork so that this average 'return period' for great CSZ quakes has a considerable amount of leeway, with some clustering also apparent; and second, the geological situation in the region is not quite as simple as previously thought. In fact, the CSZ is not one big subduction zone at all, but is instead made up of separate segments, perhaps four altogether, some of which can be expected to rupture quite soon, resulting in severe ground shaking and a major tsunami.

According to Chris Goldfinger and his collaborators, the type and size of earthquakes occurring on the CSZ depend upon where they originate. Those that start in the north seem to tear the entire 1000 km fault apart, triggering the biggest earthquakes that, at about magnitude 9.0, are comparable with the 2011 Japan event. More frequent than these are smaller, but still—at magnitude 8 or more—scarily huge, quakes that start at the southern edge of the subduction zone. Perhaps the most worrying observation of Goldfinger's group is that the time since the last CSZ earthquake is already longer than three-quarters of the known intervals between earthquakes during the past

10,000 years. By 2060, this figure will have risen to 85 per cent unless, that is, another Big One strikes in the meantime.

Notwithstanding any potential influence from rising sea levels, Goldfinger and his team have worked out the chances of the next great CSZ quake occurring in forthcoming decades. Their worrying conclusions are that the southern end of the fault has a 37 per cent chance of generating a major earthquake in the next 50 years, in other words a little over 1 in 3. The odds on a massive quake starting in the north and rupturing the entire fault are—at between 1 in 7 and 1 in 10 in the next 50 years—somewhat more favourable, but still discomforting. To add to anxieties about this worst-case situation, the northern quakes have a tendency to cluster so that for a period they may shake the region as frequently as every 250 years while, at other times, seismic tranquillity may reign for as long as a thousand years. The question that begs answering is: where in this sequence of shake-and-sleep are we now? In the context of climate change we might also enquire: could the recent or future trend in rising sea levels bring forward the arrival of the next big quake or will it extend the length of the current period of seismic quiescence? To be honest, evaluating the influence of sea level on the likelihood of a single event is going to be difficult at best, and it is likely that any relationship between changing sea levels and earthquake activity in the CSZ will only be recognized in a detailed analysis of the seismic record over a long period of time. In fact, there is indeed a hint that prior to 9000 years ago, when Holocene sea levels were shooting upwards most rapidly, the frequency of CSZ earthquakes was somewhat reduced. Although any cause-and-effect link with changing sea level remains speculative, such a finding would fit with the model proposed by Luttrell and Sandwell, the implication being that the CSZ was preferentially rupturing deeper and further to the east beneath the western edge of the North American continent, as the load due to higher sea levels stabilized the shallower levels of the fault offshore and reduced the frequency of the giant, tsunami-generating quakes such as that which shook the region in 1700.

Riding high

Notwithstanding the fact that rock-bottom global sea levels were a boon to the countless mid ocean ridge volcanoes seeking to erupt more magma, to Italy's Pantelleria volcano; and probably to other volcanic islands scattered around the ocean basins, it looks as if eruptions at many volcanoes are promoted when sea levels are riding high. As described in chapter 3, the volcanic reaction to the ameliorating climate of the post-glacial world is most obvious, and most easily explained, in regions previously covered by thick ice sheets, or close to such. However, my own team's work on the timing of eruptions at Mediterranean volcanoes points to a similarly lively rejoinder from volcanoes remote from the ice sheets but in close proximity to the oceans. To recap, the Mediterranean study, published in *Nature* in 1997, revealed a clear correlation between how quickly sea levels changed over the past 80 millennia and the frequency of important eruptions in the region. The relationship was strongest during the early Holocene from 15,000 to 8000 years ago, which saw some of the most extraordinarily rapid rises in global sea level ever recorded.

From a study of coral reefs, undertaken in the mid 1990s, Paul Blanchon and John Shaw, of the University of Alberta, Canada, were able to quantify just how fast. Utilising the principle that reefs grow vertically so as to keep up with rising sea levels, the Alberta pair were able to recognise and date three 'catastrophic rise events' or CREs, which occurred between a little over 14,000 years ago and something under 8000 years before present. The oldest surge saw sea levels rocket upwards by around 13.5 m in around 300 years; the next involved sea levels climbing 7.5 m in about 160 years, and the youngest pushed water levels up a further 6.5 m in less than 150 years. Blanchon and Shaw reckon that the rises must have been sourced, initially, by megafloods of glacial meltwater into the ocean basins. This, in turn, they suggest, destabilized ice sheets grounded below sea level, causing widespread collapses that launched fleets of icebergs into the Atlantic. On melting, the water contained in these pushed sea levels up even further, leading to a knock-on effect that destabilized ice sheet margins even more. As an aside, and a nod to the future, the Canadian researchers observe that this behaviour should provide us with a salutary warning in relation to the ultimate fates of the Greenland and West Antarctic ice sheets and the rapidity with which global sea levels could soar should these great ice repositories start to fall apart in earnest.

Over the period spanning the catastrophic sea level rises of the early Holocene, the frequency of notable eruptions at Mediterranean volcanoes leapt up from just over one thousand years to about 350 years. It would be rather odd if such a quickening of volcanic action was limited to one small sea and, indeed, the sulphate record preserved in cores extracted from the Greenland Ice Sheet points to a widespread volcanic response at this time. There are plenty of potential ways in which large and rapid rises in sea level could have triggered such a fiery outburst from volcanoes in or close to the sea and these have been examined in chapter 3. However, it is worth drawing particular attention here to what looks like being a key driver of enhanced volcanic activity at such times, as well as increasing earthquakes as noted above: the bending of the lithosphere around the margins of the ocean basins due to increasing weight of water. The work of Luttrell and Sandwell has shown that this phenomenon was capable of promoting earthquakes at coastal faults during the replenishment of the oceans during post-glacial times, while Steve McNutt and John Beavan have resorted to it to explain how seasonal changes in local sea level influence the timing of eruptions at Alaska's Pavlof volcano. If a short-term sea level rise of less than 20 cm is sufficient to squeeze magma out of Pavlof today, then prospects must be excellent for bending of the coastal lithosphere promoting a far wider volcanic response during the extreme marine inundations of the early Holocene.

So it looks as if bending of the lithosphere around the margins of replenishing ocean basins provides a common mean, both of triggering earthquakes and encouraging the rise and expulsion of magma. Accompanying phenomena may also drive hazardous outcomes, particularly in relation to decreasing the stability and promoting the failure of slopes. Opportunities for pore pressure change and slow cracking to trigger landslide formation in the coastal environment are greatly increased at times when sea levels are climbing rapidly. Instability and collapse may also be promoted, or at least aided, by rapid erosion, as water levels progressively eat away at cliff faces and the flanks of volcanoes. The stability of the latter seems to be especially under threat at marine volcanoes, with the progressive loading of adjacent seawater altering internal stresses so as to encourage sideways sliding, a particularly effective way—as Mount St Helens taught us—of triggering a violent blast. Certainly, the evidence presented in previous chapters seems to point to volcanoes having a preference for collapsing during periods when our world's climate is warm and wet and when sea levels are high or heading upwards.

Unholy trinity: landslides, tsunamis and gas hydrates

Despite evidence for some major undersea slides occurring at times when sea levels were low or falling, there seems to be little doubt that submarine slopes have a much greater tendency to fail when sea levels are high or heading upwards, a point discussed at some length in chapter 5. Given the colossal volumes of many of the submarine landslides of the Holocene, most—if not all—might be expected to generate significant tsunamis. The great Storegga Slide testifies to the tsunami-forming potential of large mass movements under the sea, as do the sandy layers in the peat of the Shetland Isles, which provide evidence of two younger tsunamis, assumed to have been sourced by submarine landslides.

On the whole, however, evidence for tsunamis caused by Holocene mass movements in the marine environment is sparse. It might be expected that tsunami deposits emplaced at times of low sea level stand a good chance of being destroyed or covered by rapidly rising oceans during post-glacial times. High sea levels should, however, provide an environment reasonably favourable for their preservation, but this does not seem to be the case; either that or we are just not looking hard enough. One lesson we have learned relatively recently is that tsunami deposits can take on many guises, from thin layers of sand to piles of giant boulders. After their formation they can also be worked over by wind and water to such a degree that their primary tsunami features become hidden. Maybe they are there after all, but we just can't recognize them. Clearly, at a time in Earth history when sea levels are high and certain to climb higher, quite possibly far higher, it is useful to know how effective submarine landslides might be at generating tsunamis. It looks, however, as if we will have to rely far more on modelling than on empirical evidence, if and when we decide we need to evaluate a future potential tsunami threat arising from the rapid deglaciation and rebound of Greenland.

Still to be resolved too is the relationship—when the oceans are warm and high—between submarine landslides and gas hydrates. As addressed previously, the breakdown of gas hydrates has been implicated by some researchers in the triggering of submarine landslides when sea levels were low, the reduced load pressures causing the dissociation of the hydrates and a massive, destabilising volume change. Of course, gas hydrates can also be encouraged to break down due to rising temperatures and invariably, replete oceans are also warm ones. As examined in some detail in chapter 2, the relationship between submarine landslides and gas hydrates remains enigmatic, and while circumstantial evidence suggests a connection between the two, the nature of the link is far from clear. In theory at least, the potential exists for balmy ocean temperatures to trigger the large-scale release of marine gas hydrates, once the warmth has penetrated into the ocean depths and into the sediments hosting the hydrate. According to some, this could lead to increased instability around the margins of the ocean basins and to a burst of submarine landslide activity, which could in turn see more tsunamis careering across the world's oceans. Of course, such a scenario depends entirely on rising ocean temperatures trumping rising ocean levels, which act to stabilise gas hydrates. The alternative argument, that a surge in the formation of undersea mass movements, occurring in response to rising sea levels, can trigger the widespread release of methane from gas hydrates depressurized as a consequence, also has its advocates—but its detractors too. The association between the two remains unclear, which is a shame as this would be something well worth knowing as sea levels continue to rise, oceans get ever-warmer, and the future of the surviving polar ice sheets becomes progressively more insecure.

To conclude, the role of the oceans in teasing out a reaction from the geosphere at times of major climate transition cannot be over-played. The response driven by the melting ice sheets as active faults and volcanoes, held captive for tens of thousands of year, are set free and vast quantities of sediments shed from northern continental margins, to send tsunamis skimming across the rising oceans is huge, but it is also spatially confined. The wholesale change to our planet's spin characteristics and the pattern of stress and strain in its interior, which accompany the filling—and emptying—of the ocean basins, transmits the geospheric response far and wide, well beyond the direct influence of the ice sheets. Because of this, the switch from glacial to interglacial conditions was able to promote earthquakes in California and volcanic eruptions around the shores of the Mediterranean. Because of this, future melting of the Greenland and West Antarctic ice sheets, and the resulting 10 m plus surge in global sea levels, might well be sufficient to drive a hazardous response from the crust that has the whole world in its sights.

7

Reawakening
the Giant

It should be clear by now that when we look to the immediate future of our planet, we also look back to its geologically recent past, in the sense that the world of the 21st century and beyond is projected to have much in common with the period of climate chaos that followed hard on the heels of the Last Glacial Maximum. That the extraordinary transformation in the Earth's climate between 20,000 and around 5000 years ago drew a lively response from the solid Earth is incontrovertible. Only recently, however, has the possibility that anthropogenic climate change may evoke a response along the same lines begun to attract attention. Research in this area is very new, but it is critical to determining whether or not our children and generations yet unborn will be faced with a picture of geological upheaval to compound all the other woes forecast to accompany remorseless anthropogenic warming. Combining what we have learned from the past with what current observations and new projections are telling us, this chapter will look forward rather than back, addressing the likelihood and nature of the prospective geological response to contemporary climate change.

Reawakening the giant

The giant woke with a vengeance on the last occasion that our planet experienced wholesale ice-mass loss and ocean levels rose to compensate, but has slumbered since. Can we be certain that the impact of human activities on today's climate will be sufficient to reawaken it, and if so, will this really make much difference? After all, one or other volcano is always erupting; a large quake shakes the planet every two or three days; and tsunamis, landslides, and mudflows are nothing out of the ordinary. Even if we fail miserably in our attempts to reduce greenhouse gas emissions so that, ultimately, we lose the Greenland and West Antarctic ice sheets and sea levels surge far higher as a result, will this be enough to elicit a reaction from the geosphere sufficiently large as to be distinguishable from everyday geological activity? My guess, for what it's worth, is that it will.

Clearly, the Earth of the early 21st century bears little resemblance to the ice world of 20,000 years ago. Today, there are no great continental ice sheets to dispose of, while the ocean basins are already pretty much topped up. On the other hand, climate projections repeatedly support the thesis that global average temperatures could quite easily rise more rapidly in the course of the next century or so than during post-glacial times, reaching levels at high latitudes capable of driving catastrophic break-up of polar ice sheets as thick as those that once covered much of Europe and North America. Additionally, even with sea levels already 130 m higher than at the peak of the last glaciation, there is still plenty of ice around, enough in fact to add a further 70 m or so in the barely imaginable, and one hopes unlikely, situation that it all ultimately melts. More realistically—in the shorter term at least—the magnitude of a future sea-level hike accompanying unmitigated anthropogenic greenhouse gas emissions is forecast to be on the same order as the catastrophic rise events of the latest Pleistocene and early Holocene. These, remember, saw sea levels shoot upwards by 4–5 m a century over periods as short as a few hundred years.

Bearing this in mind, recent disturbing evidence from the last interglacial period suggests that the polar ice sheets may be far more vulnerable to even a small degree of sustained global warming than previously thought. Princeton climate scientist, Robert Kopp, along with colleagues from the same university and others at Harvard, revealed in a 2009 paper in *Nature* that despite temperatures at high latitudes being just 3–5°C warmer than today, during the Eemian, there was a 95 per cent probability that global sea levels were at least 6.6 m higher, and a 67 per cent chance that they were elevated by 8 m compared to the early 21st century. Kopp and his team also attempted to estimate how quickly sea level is likely to have risen as the Eemian climate warmed, coming up with a range of 5.6–9.2 m in in a thousand years. This translates, on average, to 56–92 cm per 100 years—close to a number of model-based estimates for the rate of sea-level rise this century, and similar to those predicted by the most recent observations of the rate of melting at the poles.

The rates at which future temperatures rise, ice sheets dissolve and sea levels climb may well, then, be at least comparable with those of post-glacial times, making the expectation of a reaction from our planet's crust not only plausible but likely. It would be logical to

Fig 24. Should all the ice on the planet melt, global sea levels would climb by 70 m or so. The implications for the UK would be the inundation of much of the lowlands that dominate the south and east of the country and the formation of an archipelago of countless small islands.

expect that the scale and extent of such a reaction would, broadly speaking, march in step with rising temperatures, but this might well be a dangerous oversimplification. Rather than a steady linear increase in the geological response, it may proceed in sudden leaps and jumps as critical thresholds are crossed and tipping points exceeded. Whatever the nature of the reaction, there will clearly be disparities in comparison to the post-glacial period; for example any volcanic riposte to the future loss of Iceland's Vatnajökull Ice Cap will be far less impressive than the extraordinary eruptive outburst that greeted the wholesale melting of the vast ice mass that buried Iceland during glacial times.

If the scale and speed of future anthropogenic climate change are great enough, however, the broad picture may well be the same, with the geosphere reacting most obviously in those regions across which wasting of ice sheets and glaciers, climbing sea levels, and increased precipitation are most pronounced. These include all the usual suspects for which evidence exists of past responses to a rapidly changing climate: the margins of the oceans, glaciated regions at the poles and at high altitudes and elevated topography—both volcanic and non-volcanic. The bottom line is that our knowledge of how the world was transformed at the end of the last glacial period reveals the all-encompassing and all-pervasive nature of rapid and severe climate change, such that the planet in its entirety is mobilized. So complex and entangled is the Earth System that, looking to the future, nothing can be regarded as immune to the influence of anthropogenic warming. Not only this, but we are already seeing the first signs of the geosphere responding to changes wrought by rising temperatures, while ongoing research suggests that there could be far more to come.

In chapter 4, I looked at the evidence for reduced load pressures resulting from widespread melting of ice cover promoting increased earthquake activity in southern Alaska. In a sense, this detached US state is acting out the role of the canary in the cage, the coincidence of plenty of ice and a geologically dynamic setting providing opportunity for early interaction between the two as the local climate warms considerably faster than the world as a whole. Alaska is predicted to be more affected by climate change than any other US state. Taking a wider perspective, what is happening in the icy wastes of Alaska today might well be providing us with a taste of what to expect in other comparable parts of the planet in decades to come—and not just more earthquakes, but more landslides too. As the wholesale melting of great ice masses such as the Bagley Ice Field free up active faults beneath, so the progressive loss of mountain permafrost and ice is destabilising rock faces and promoting the formation of giant landslides.

And every mountain and hill shall be laid low

Alaska is home to impressive peaks of both a volcanic and a non-volcanic disposition, including the currently inactive Mount Bona, which rises over 5000 metres, and Mount McKinley, at 6198 m the highest mountain in the USA. A combination of very high and very steep terrain and a tectonically active environment means that impressively big landslides are not uncommon. In recent years, however, they seem to be happening far more frequently. On the 3000 m Iliamna volcano, huge rock and ice avalanches with volumes of 10–30 million cubic metres (about enough to fill 4000 to 12,000 Olympic-size swimming pools) used to happen only about once a decade. During the past 15 years, however, their frequency has increased dramatically to once every three to five years. This could simply be a reflection of the fact that no-one was paying sufficient attention until more recently, but for the fact that the volcano is not alone in regularly divesting itself of very large volumes of rock and ice. In 2005, a glacier capping the 3200 m peak of Mount Steller, in southern Alaska, detached itself and plummeted down the flanks of the mountain in the form of a 50 million cubic metre maelstrom. The avalanche is likely to have travelled at a speed of 100 metres a second (360 kph) or perhaps even faster, driving the collapsing mass for a distance of nine kilometres before it came to rest on the surface of the Bering Glacier below. So violent was the event that it was recorded on seismometers around the world.

Along with a number of other scientists, Christian Huggel of Zurich's esteemed technical university, Eidgenössische Technische Hochschule (ETH), and Jacqueline Caplan-Auerbach of the Western Washington University, in the eponymous state, have been keeping a special eye on Alaska's recent landslide activity. Their studies suggest that at Iliamna, the driving force for repeated landslides is likely to be volcanic heat, which provides an efficient means of melting the base of the attached glacier, promoting its periodic collapse. They point out that climate plays a role too by providing the raw material that replenishes the glacier following its failure. This takes the form of a good 10 m of snow a year that falls on the volcano. One could speculate that more frequent failure could be a reflection of the more rapid restocking of glacier mass, perhaps as a consequence of increased precipitation due to the impact of climate change in the region. The state is progressively becoming wetter, with precipitation increases of up to one-third measured across much of Alaska between the late 1960s and early 1990s.

Volcanic heat cannot be blamed for the giant Mount Steller avalanche of 2005, and here a potential role for climate change seems to be more explicit. Huggel, Caplan-Auerbach, and their colleagues point to the fact that global warming is pushing up temperatures at a rate of 0.3–0.4°C a decade at the height of Mount Steller's summit. They also observe, perhaps more significantly, that summer temperatures in the region were unusually high during both 2004 and 2005 (the avalanche occurred in September 2005) heating up the atmosphere around the mountain and reinforcing the year-on-year temperature rise. According to the watching scientists, such a combination could have driven sufficient melting of glacier ice and thawing of mountain permafrost to trigger failure and avalanche formation.

Alaska is far from the only part of the world to have hosted giant landslides in recent decades, and their apparently increased frequency in many mountainous regions seems likely to represent the first extended response of the geosphere, at a global level, to anthropogenic climate change. Half a world away from the state of grizzlies and moose, 2002 saw the triggering of the biggest rock and ice avalanche so far observed, high up in Russia's Caucusus Mountains, sandwiched between the Black Sea to the west and the Caspian to the east. On a chill evening in late September, a huge chunk of rock and ice clinging to the summit of Dzhimarai-khokh detached itself, sending somewhere between 10 and 20 million tonnes of debris crashing onto the Kolka Glacier far below. The immense force sliced off most of the glacier, incorporating its ice into the avalanche that continued to hurtle downslope. An estimated 100 million cubic metres of rock and ice cascaded into the Genaldon Valley and burst upon a small, sheltered, bowl between two ridges, known as the Karmadon Depression. Those living in the farming communities within had no time to escape and 120 men, women and children now lie entombed in their homes beneath a mountain of debris. After travelling 19 km, the bulk of the avalanche was halted at a narrow defile known as the Gates of Karmadon, where giant blocks of ice and rock became wedged, eventually forming an immense natural dam. Even so, the watery residue of the landslide managed to force its way through, feeding a muddy torrent that continued for a further 15 km, stopping just short of the much larger town of Gisel. Although climate records for the Caucasus region are poor, the rapid retreat of glaciers, which has accelerated since the 1990s, attests to a general and continued warming, which can reasonably be assumed to have played a role in triggering the initial avalanche.

In other parts of the world too, previously enduring mountain faces are tumbling. The Monte Rosa peak in the Italian Alps has sourced frequent ice and rock avalanches since the 1990s, culminating in the removal of more than a million cubic metres of ice and debris from its east face in 2005. Elsewhere in the Alps, major collapses—each involving a million cubic metres or more of material—have been documented at Monte Bianco (Mont Blanc) in 1997 and Punte Thurwieser in 2004, both in Italy, and at the Swiss peaks of Dents du Midi and Dents Blanches in 2006. Further afield, recent large landslides and avalanches have also been reported from Canada and from the Alpine backbone of New Zealand.

It is fair to expect that as the world continues to heat up, we will see a steady ramping up in the number of large and potentially catastrophic landslides, at least until all the ice and mountain permafrost has gone. Instead, however, landslide activity might be promoted, as it seems to have been at Alaska's Mount Steller, during or following years in which the summers were particularly hot, resulting in a more erratic threat. This is something that has also attracted the interest of Christian Huggel and others, who have looked for, and found, clear links between the timing of major rock and ice avalanches in recent decades and the occurrence of heat waves. In every case that was evaluated—at Mount Cook, in New Zealand's Southern Alps, in 1997; Mount Steller in 2005 and 2008; Mount Miller, also in Alaska, in 2008; and Monte Rosa in 2007—collapse was heralded by unusually warm weather. This is not good news. As mentioned in chapter 1, projected future rises in global average temperature hide much of the detail, with the land warming faster than the oceans and the polar regions heating up fastest of all. In addition, periods of extreme temperature are forecast to rise twice as quickly as average temperatures and to last longer, making blistering heat waves far more common. In Europe, the baking summer of 2003 will likely be regarded as run-of-the-mill by 2040 and—20 years later—as pleasantly cool. This is not ideal from the perspective of encouraging increasingly rickety faces of Alpine peaks to stay put. The same pattern is likely to be repeated in glaciated mountain regions across the world, from the peaks of Alaska to the Himalayas, and from the Andes to the Southern Alps, with longer and hotter heat waves reinforcing the trend towards increasing instability of mountain faces and glaciers driven by a general, year-on-year, warming.

With the exception of the Caucasus event, most recent major landslides have occurred far from inhabited areas. As the world continues to warm, however, and bouts of extreme temperature become more common, future collapses are bound to impinge upon areas where population densities are far higher. In the European Alps, where a landslide in 1881 destroyed the Swiss village of Elm, mountain communities will inevitably find themselves increasingly at the mercy of mass movements as the ice and permafrost holding the surrounding peaks together succumbs to higher air temperatures. In the Andes, where 18,000 died in 1970, obliterated by a massive rock and ice avalanche triggered by an offshore earthquake, progressively rising temperatures are likely to seriously increase the landslide risk to those towns and villages huddled beneath the world's longest land-hosted mountain range.

Notwithstanding their capacity to wipe entire communities from the face of the planet, more landslides will also bring another threat. In glaciated mountain regions where ice is melting fast, large masses of collapsed debris make very effective natural dams capable of impounding huge volumes of meltwater. Failures of such dams, perhaps as a consequence of overtopping or breaching by an earthquake, can source catastrophic floods powerful enough to travel downslope for many tens, if not hundreds, of kilometres, bringing the threat of inundation to major urban centres far removed from the original event. While attributed to an earthquake rather than a changing climate, a giant landslide in the Central Asian republic of Tajikistan, 100 years ago, not only highlights this potential problem, but—in its own right—continues to present a major threat today.

Lake Sarez—a Damoclean sword in Central Asia

In February 1911, shaking due to a moderate earthquake detached a colossal 2.2 cubic kilometres of rock from the side of a steep-sided valley through which flowed the Murghab River. The fact that the slide buried the village of Usoi, incarcerating all but four of its population of 300 or so, is almost incidental in respect of what happened next and what may well happen in the future. The landslide came to rest across the valley forming—at 5 km in length and 567 m high—the world's tallest natural dam, named the Usoi Dam after the obliterated village. Inevitably, with nowhere else to go, the river waters rose steadily behind the new barrier, quickly flooding the neighbouring village of Sarez and fashioning an immense lake that today is backed up for more than 55 km behind the rock dam and which contains 16 cubic kilometres of water. Because of the threat to densely populated areas further down the Murghab valley, worries about the stability and permanence of the rock dam burgeoned within months of its emplacement. Concern increased in 1968 when a small landslide sent miniature tsunamis across the lake, and again in 1987 when a bigger collapse into the lake spawned 6 m waves; the threat of a future catastrophe arising from the breaking of the dam continues to focus minds.

Two scenarios are under the microscope: the first involves the collapse of a large mass of unstable rock above the lake—perhaps due to another big earthquake—generating tsunamis that overtop and erode the dam, ultimately releasing the lake waters into the network of river valleys below, which together host a population of more than five million people. The second envisages a future quake shaking the dam itself sufficiently violently to allow the lake to break through. A number of studies have come to conflicting conclusions about the stability of the dam and the probability of a future collapse but this is not a problem that looks like going away. Water levels are rising faster as the 900 or so glaciers capping surrounding peaks melt. Seepage through the dam is increasing and the threat from earthquakes is ever present, both offering means by which further landslides could be triggered that could destabilize or breach the dam. Incredible as it sounds, a 1998 model developed by the US Army Corps of Engineers predicts that—after blasting through a maze of narrow, densely populated valleys—a worst-case flood crest would still be as high as a two-storey house when it reached the town of Termiz (population 140,000) on the border between Uzbekistan and Afghanistan, 1400 km downstream of Sarez.

Natural dams similar to that at Sarez abound in mountainous regions, most especially in the Andes and Himalayas, holding back bodies of water that are growing daily as warmer temperatures accelerate melting of glaciers and ice caps. In the Peruvian Andes alone, more than 30 so-called glacial outburst floods—or GLOFs—have together taken 6000 lives since the early 1940s. In 2010, an avalanche of ice and rock from the Hualcán Glacier plunged into a Peruvian lake, giving birth to a 25 m tsunami that flooded four towns and destroyed 50 homes. Across Patagonia too—whose ice fields make up the third largest ice mass on Earth, after the Greenland and Antarctica ice sheets—glacial meltwater lakes are grow-

Fig 25. Lake Sarez, in the Central Asian Republic of Tajikistan, was impounded by a natural rock dam formed by a colossal landslide triggered by an earthquake in 1911. Should the dam fail, the 16 cubic kilometres of water making up the lake will present a massive threat to densely populated areas further down the Murghab valley.

ing steadily and increasing in number as ice in the high Andes melts at ever faster rates. Meanwhile, the International Centre for Integrated Mountain Development has pinpointed more than 200 glacial lakes in India, China, Pakistan, Nepal, and Bhutan, which threaten to burst through confining dams of rock or debris, pouring torrents of icy meltwater onto unsuspecting communities in valleys far below.

As the world continues to warm and mountain glaciers shrink at ever faster rates so far more meltwater lakes will appear, while those already in existence will expand and deepen. This is likely to raise both the likelihood of outburst floods and their destructive potential. Rising water levels will provide conditions more conducive to slow cracking, the process that triggered the collapse of the north flank of

Mount Toc into the Vajont reservoir in 1963. Similarly, pore water pressurisation may encourage movement of adjacent faults, raising the possibility that the largest lakes, at least, might promote local earthquakes. The resulting shaking would, in turn, have the potential to breach the offending lake at its weakest point, triggering the catastrophic release of impounded waters. In the seismically active Andes and particularly in the Himalayas, where half a dozen great earthquakes are due, tectonically sourced violent ground-shaking may do the job even more effectively.

While the growing threat of large-scale mass movements in glaciated high mountains is nourished by escalating temperatures, it is worth remembering that anthropogenic climate change will see many parts of the world getting wetter as well as warmer—indeed, observational evidence suggests that this is already happening. More intense rainfall events, in particular, can be expected to drive an increase in the formation of debris flows, as well as promoting landsliding through the saturation of slopes and the pressurisation of pore waters. Areas especially susceptible to the mobilisation of unconsolidated surface material by heavy rains include bare surfaces recently exposed by retreating glaciers, the ash-covered slopes of volcanoes and steep terrain that has been deforested by logging.

One foretaste of what we might expect is supplied by the collapse of part of Nicaragua's Casita volcano in 1998 as a consequence of torrential precipitation across Central America produced by Hurricane Mitch. This led to the formation of deadly mudslides that snuffed out, in an instant, more than 2000 lives. Another is provided by the fast-flowing torrents of mud and debris caused by an intense downpour falling on loose volcanic soils which, in the same year, destroyed or damaged 600 homes and killed more than 160 people in the southern Italian town of Sarno. More recently, in 2005, a massive debris flow blocked the Aare valley in the Swiss Alps, temporarily damming the river Aare, which eventually burst through the dam causing serious flooding. The flow was initiated by heavy rains falling on loose ground-up rock, pulverized by the now retreating Homan glacier in the mountains above the valley. With a volume of half a million cubic metres, the flow was the largest in the Alps for at least the past two decades. Five years later, and on another continent, torrential rains triggered more than 40 landslides, blocking all routes linking the iconic Peruvian Inca settlement of Machu Picchu with the outside world and requiring 2000 tourists to be airlifted to safety. With tropical cyclones forecast to get wetter, and intense downpours expected to increase generally, it is hard to think of a reason why destructive and lethal mass movements driven by precipitation should not become increasingly apparent.

A fiery future?

It would be odd if any reasonably minded person thought that a world of higher temperatures, melting ice and more intense rainfall would not also be one in which slopes, peaks, and rock faces became more unstable and the downslope transport of rock and debris more common, but what about potentially hazardous geological phenomena driven by deeper, more fundamental processes? Will we see, as in post-glacial times, a more fundamental riposte to anthropogenic climate change in the form of more volcanic outbursts and earthquake shocks, and if so when? The answers depend upon many things, but most critically upon what action we take in the next few years to seriously tackle climate change. If, by means of slashing global greenhouse emissions, we can prevent large-scale melting of the polar ice sheets and the consequent major hike in global sea levels, then the response from the crust may be subdued; the giant beneath our feet stirring a little, as it were, before returning to his slumbers. The problem is that we have so little time to act. Realistically then, it may be too late to stave off the catastrophic loss of polar ice and rocketing sea levels. The glaciers of Greenland are already retreating en masse, dumping hundreds of billions of tonnes of meltwater into the Arctic Ocean every year and accounting, even now, for one quarter of sea-level rise. But this is not the complete picture. In Greenland and across the planet, glacier melting is always lagging behind climbing temperatures, so that even if the world stops heating up today, melting will continue. In 2010, more than half of the surface of the Greenland Ice Sheet experienced melting, and prospects for its survival look very bleak indeed. According to a model developed by Sebastian Mernild, of New Mexico's Los Alamos National Laboratory, and his co-researchers, the measures we are taking now to save Greenland are simply not enough. Even if we embrace renewable technologies, continued population rise and economic growth will doom the ice sheet. Mernild's team reckon that Greenland will reach a tipping point in about 30 years' time, after which there is no turning back. Nothing we do, thereafter, will save the ice sheet from extinction.

It would be nice to be proved wrong, but based upon such predictions, alongside the absence of any serious attempts to cut global greenhouse gas emissions in line with the findings of climate scientists, it has to be assumed that during the course of the 21st century and beyond, temperatures will rise, ice sheets and glaciers melt, and ocean levels climb, at rates that are comparable to those in post-glacial times. In such circumstances, it would be surprising indeed if there were not a distinguishable and measureable reaction from the Earth's crust. Recognising this, a number of studies are zeroing in on those parts of the world where any tectonic response might be first and most easily detected. Inevitably, Alaska, where a seismic reaction to melting glaciers is already apparent, is at the top of the list. Near the top, too, is Iceland, where the volcanic rejoinder to the loss of the

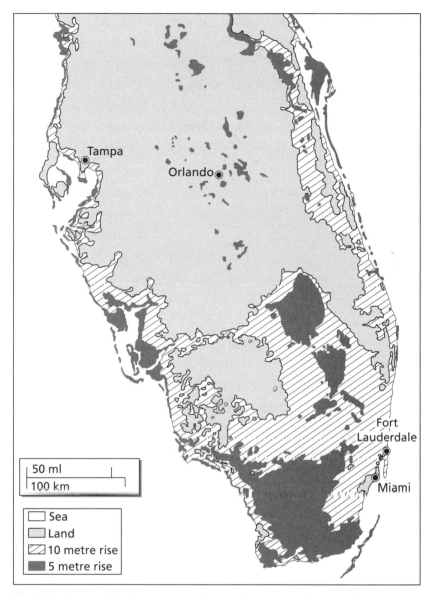

Fig 26. It is possible that in about 30 years' time we will reach a tipping point that makes the loss of the Greenland Ice Sheet certain, whatever action we take thereafter. The resulting 7m rise would swamp low lying parts of the world such as southern Florida.

countrywide ice cover at the end of the last glaciation was spectacular. Looking to the future of volcanic activity in this land of fire and ice, attention is focused on the remaining glaciated areas, by far the greatest of which is the Vatnajökull Ice Cap, host to the devastating Lakagígar eruption of 1783.

With prospects for the survival of Vatnajökull in an anthropogenically warmed world seemingly very poor, Icelandic volcanologist, Freysteinn Sigmundsson, and his team, have looked closely at what this might mean for the volcanoes that are currently residing beneath it. While tiny in comparison to the great ice sheet that buried the island 20,000 years ago, the Vatnajökull Ice Cap is still impressive. Covering an area of around 8000 square kilometres, it is about the same size as urban New York, and in places is still a kilometre thick. Already, however, it is on the wane. Between 1890 and 2003 the ice cap lost about one tenth of its mass—about 400 billion cubic metres in all—and it is now getting half a metre thinner every year. To compensate, the lithosphere is starting to bounce back, uplifting by a finger or two's width every year across much of its extent. This in turn is reducing pressures in the asthenosphere— the magma nursery beneath—making melting here easier. A repeat of the astonishing volcanic revival that greeted the end of the last ice age, when the production of magma beneath the island shot up 30-fold or more, is completely out of the frame. Nevertheless, on the basis of their calculations, Sigmundsson and his colleagues do expect a more measured volcanic revival as the ice cap gradually crumbles away to nothing. They estimate that the reduced ice load will lead to another billion and a half cubic metres of magma being produced in the asthenosphere every century—not a vast increase, but sufficient to supply a moderate eruption every 30 years or so. Probably best not to hold your breath, however; Sigmundsson and his co-workers speculate that it may take centuries for the new magma to reach the surface. As the ice cap continues to melt and load pressures on the plumbing systems of the volcanoes beneath are progressively reduced, a more rapid response may come from the early release of magma from volcanoes that are already building towards eruption, in this way increasing the frequency of volcanic events. More volcanic outbursts would clearly not be a good thing for Iceland and its inhabitants, and probably not for the North Atlantic region as a whole. After all, we only have to go back to 2010 and 2011 to get an idea of the future potential for disruption of even quite small volcanic flare-ups in Iceland's Vatnajökull region. A better impression of the longer-term threat presented by any increase in the number of Icelandic eruptions is, however, provided by the 1783 Lakagígar event, the knock-on climatic effects of which may have taken millions of lives across the world. A similar type of eruption today would likely cause additional mayhem by

Fig 27. The 2011 eruption of Iceland's Grimsvötn volcano: could the eventual loss of the Vatnajökull Ice Cap bring about a rise in eruptive activity at volcanoes previously imprisoned beneath the ice?

closing down traffic along the North Polar air routes for months at a time.

While Iceland is a prime candidate to host a volcanic response to a warmer climate, it is far from the only one. Ice-covered volcanoes abound in many parts of the world, most especially in Alaska and down the west coast of the USA, in Russia's Kamchatka Peninsula, and along the length of the Andes. As climbing temperatures divest these volcanoes of their icy carapaces, so the potential for more violent blasts is increased. According to University of Lancaster volcanologist Hugh Tuffen, the thinning of ice cover may not only drive more explosive eruptions as pressure is released on stored magma beneath, but the removal of mechanical support provided by previously permanent ice cover may make a volcano more prone to lateral collapse, such an event being capable in its own right of triggering a violent eruption. At the very least, the fields of ash and loose debris exposed as ice is progressively lost will provide a ready supply of material for debris flows, triggered either by meltwater or intense rainfall. Inevitably, more rock and ice avalanches are to be expected as volcanic heat is able to more easily dislodge masses of melting and degraded ice weakened by higher air temperatures, and glacial outburst floods may be a problem too, these arising from the catastrophic release of meltwater from the overtopping or breaching of crater lakes or depressions, particularly where the water is impounded by weak ash and rubbly debris. Such an event sent one and a half million cubic metres of water and mud cascading down the flanks of New Zealand's Ruapehu volcano in 2007, when a debris 'dam' holding back a crater lake was breached. Looking ahead, a response can also be expected at volcanoes abiding at low latitudes or low elevations, which have never hosted so much as a single ice crystal, but which are forecast to experience a wetter climate as global temperatures rise. In the Caribbean, Europe, Indonesia, the Philippines, and

Japan, more intense rainfall episodes—linked in the tropics to wetter hurricanes and typhoons—are likely to increase prospects for the formation of potentially destructive debris flows and encourage landsliding. The Casita tragedy, touched upon earlier, could quite reasonably be regarded as a forerunner of more such events.

Notwithstanding a rise in the number of volcanic debris flows and relatively minor—though still potentially lethal—landslides, prospects for an increase in prodigious lateral collapses at the world's volcanoes must be a worry as temperatures climb, particularly in light of evidence suggesting that volcanoes seemingly have a propensity for falling apart when the Earth's climate is relatively warm and wet. As mentioned above, loss of stabilising ice cover may provide one possible trigger, but there are others. Saturation of the upper levels of a volcano, either due to elevated levels of meltwater production or to a wetter, local climate, can pump up pore water pressures, so raising the potential for sliding along existing planes of weakness. As seems to have happened at Sicily's Mount Etna during the early Holocene, the intrusion of fresh magma into a soggy volcano also provides an excellent means of provoking collapse of the flanks, and perhaps—as an added bonus—an eruption. Sea level too, may have a role to play, especially if—as seems perfectly feasible—we are in for a rise of several metres in coming centuries. The resulting bending of the crust in coastal areas that host volcanoes, discussed previously, acts to promote instability on the seaward flanks, as well as easing the ascent and extrusion of magma, should it be readily available at shallow depths. This latter behaviour—most ably demonstrated by Alaska's Pavlof volcano—brings us to a key point in relation to the likely scale and extent of a future volcanic response to planetary warming. Will this be limited to formerly ice-covered volcanoes and those subjected to significantly wetter climates, or—with 800 or so active and potentially active volcanoes forming islands or having coastal locations—will modification of stress and strain conditions in the crust, brought about by a large and rapid rise in sea level, drive a far wider response?

Between 12,000 and 7000 years ago, it is estimated that volcanic activity on land rocketed by between two and six times, an extraordinary leap that we would soon notice if it happened today. In a normal year, 50 volcanoes may erupt—some doing so almost continuously, others for the first time in decades, centuries or even longer. Imagine if this figure was upped to 100 or 300. But are we really likely see such a jump in activity levels? Arguing against such an idea is the fact that any increased volcanic action in Iceland will be far below that which enlivened the immediate post-glacial period. Similarly, with many more volcanoes at high and mid latitudes supporting far thicker covers of ice during the late Pleistocene than they do now, a future response here is also likely to be less palpable. Volcanoes located at the coast, or that form volcanic islands, would also, surely, not be reinvigorated by even a 10 m sea level rise in the same way that they would by one 13 times greater. Or would they? The example of Pavlof shows that provided a volcano is primed, even a relatively minuscule change in sea level can be sufficient to trigger the release of magma. More broadly, the recognition of a seasonal pattern in volcanic activity caused by tiny deformations driven by the annual perambulation of water around the planet, also underscores the extreme sensitivity of many volcanoes to negligible changes in their environment. This fact is reinforced by evidence for activity at individual volcanoes being influenced by storms, heavy rainfall or ups and downs in atmospheric pressure.

Should global sea level climb by 1–2 m this century, the rate of increase will be 25–40 per cent as rapid as in the catastrophic sea level rises between 15,000 and 7000 years ago, during which time the volcanic response to the end of the last ice age was going full tilt. Undoubtedly, part of this response would have been driven by the fading ice cover across Iceland and other volcanic regions, which was still continuing at this time, but evidence from the Mediterranean described in chapters 3 and 6 points to rising sea levels playing a key role in this pinnacle of volcanic action. Given anything like a comparable rise, it would be something of a surprise if the consequent loading of the crust by the encroaching ocean did not stimulate some degree of increase in volcanic activity later this century and beyond.

Research published in 2011 adds credence to the idea that sea levels will climb much more rapidly than predicted in the IPCC's Fourth Assessment Report and highlights an astonishing and disturbing acceleration in the rate of melting at both ends of the world in recent years. Eric Rignot of the University of California, Irvine, and his colleagues, reveal that every year, the Greenland and Antarctic ice sheets together, are now losing an additional 36 billion tonnes of mass, which is being offloaded into the oceans. In 2008, the combined mass of ice melted in Greenland and Antarctica is estimated at a staggering half a trillion tonnes. Assuming melting continues to accelerate at current levels, Rignot and his team forecast that global sea levels will be a full one-third of a metre higher by 2050. Fifty years later and a continuation of this rate of ice mass loss will ensure that melting of the Greenland and Antarctic ice sheets alone will raise sea level by more than half a metre. To this must be added further rises due to thermal expansion of the oceans as they continue to heat up, plus contributions from the thousands of smaller glaciers and ice caps that are in full-scale retreat around the world, which together would take the rise over the one metre mark. Of course, it is perfectly possible—even likely—that the rate of acceleration of ice mass loss at the poles will increase, perhaps quite dramatically, as temperatures at high latitudes climb far more quickly than the global average, in which case, who knows how high ocean levels will have risen when Big Ben rings in the new century.

Should future rocketing sea levels provoke an obvious volcanic rejoinder, an important caveat needs to be attached. The release of more magma at the surface does not necessarily reflect the production of more magma deep down in the asthenosphere. Today's annual volcano season can only be explained if the timing of eruptions is modulated or adjusted by changing environmental conditions, so that eruptions that were going to happen anyway are encouraged by favourable circumstances at certain times of the year to cluster, rather than be distributed randomly over a 12-month period. In the future, any response of coastal and island volcanoes to stress and strain variations promulgated by rising ocean levels would be in the same vein, the timing of eruptions at various volcanoes being brought into line by external 'forcing' so that over a given time frame, the level of global volcanic activity demonstrated an increase.

At the same time as this provides us with some inkling of what to expect on the volcanic front as the grip exerted by anthropogenic warming tightens, it also raises a number of key questions. Will a rise in the number of erupting volcanoes be great enough to be distinguishable above the background rate or will it involve an increase so small as to be lost in the statistical 'noise'? Will any intensification be linear, so that we see progressively more volcanic events decade-on-decade, or will it proceed in steps or jumps as certain critical thresholds are crossed? Probably most importantly, will any future response to rising sea levels be limited to especially sensitive volcanoes that erupt easily and frequently, but whose eruptions are small to moderate, or will dormant volcanoes that source far larger explosive blasts also feature? This is important, not only because such eruptions have the potential to be far more destructive and life-threatening, but also for the reason that—like the 1815 explosion of Tambora—they can have a noticeable cooling influence on the climate. Barring another super-eruption on the scale of the prehistoric Toba blast, no volcanic event is going to stop anthropogenic climate change in its tracks; another Tambora or two in a short space of time could, however, provide us with a breathing space—at least for a year or two.

Shake, rattle and roll

If melting ice and climbing sea levels prove capable of promoting an increase in volcanic outbursts, what are the prospects for active faults joining in? In a hothouse Earth of the future will more earthquakes simply be another sign of the times? As mentioned earlier, talk of a role for climate change in recent earthquake activity refuses to go away. This idea has been fed partly by the unprecedented cluster of massive earthquakes that have taken their toll across the world in the past decade, both attracting the notice of the general public and taxing the minds of Earth scientists. Since 1900, seven quakes have scored 8.8 or higher on the Moment Magnitude Scale. Of these, the first half of the 20th century saw just one, which rocked Ecuador in 1906. The following 50 years saw three such megaquakes: one beneath Russia's Kamchatka Peninsula in 1952, another off the coast of Alaska in 1964, and in 1960—registering a magnitude of 9.5—the greatest earthquake ever recorded, which tore open the Peru–Chile Subduction Zone. With the first decade of the millennium not far behind us, however, we have already seen three huge earthquakes, of magnitude 8.8 or greater, bringing death and destruction to the Indian Ocean, Chile, and Japan, all occurring within the space of less than six and a half years. Is this simply a statistical fluke? After all, the Earth does not operate to a timetable so such clustering is perfectly possible. Alternatively, is some underlying mechanism leading to a number of great quakes coinciding closely in time? We know for certain that an earthquake occurring on one part of a fault can put pressure on the next bit along—a process known as stress transfer—which can then bring forward the timing of the next earthquake here. In this way, for example, Turkey's North Anatolian Fault 'unzips' from east to west over a century or so, triggering a series of earthquakes in succession. In the context of the latest seismic cluster, however, the crucial question is: can a giant earthquake in one part of the world cause others many thousands of kilometres away and years into the future? At least one US seismologist has speculated that the huge quakes that struck Chile in 2010 and Japan the following year, might be thought of as 'aftershocks' of the great Sumatran earthquake of 2004. At present this remains pure conjecture, but in time it may be an idea that attracts further interest and research. If an active fault is critically poised and ready to rupture, as the fault off the north-east coast of Japan undoubtedly was in March 2011 after a thousand years or so of repose, then perhaps the tiny added stress needed to set it off might be provided by another massive quake several years earlier and a few thousand kilometres away. John McCloskey, of Northern Ireland's University of Ulster, has talked of a major quake in certain circumstances, needing nothing more than 'the pressure of a handshake' to trigger it, so maybe the idea is not so far out after all. As with primed volcanoes, it seems that earthquake faults that are right on the edge need very little encouragement to send them over it.

This idea leads naturally into contemporary climate change, and its potential for provoking increased earthquake activity in a warmer world, where load changes due to melting glaciers and rising sea levels will act to modify the stresses prevailing on at least some active faults. Notwithstanding the fact that the ailing glaciers of Alaska are already encouraging more earthquakes, there are other parts of the world too where a seismic reaction may become apparent as ice mass is slashed, for example in the Himalayas, Andes, New Zealand's

Southern Alps, and beneath eastern Iceland's Vatnajökull Ice Cap. Across such regions quakes may arise, as in Alaska, as a consequence of reducing the load on active faults whose movements are currently hindered by overlying ice. Earthquakes may also be induced in other ways, such as through increases in pore-water pressure in the vicinity of adjacent faults driven by growing and deepening glacial meltwater lakes, and also as a result of sudden reductions in load on faults beneath by the catastrophic draining of such lakes.

Of most interest, however, is any potential for climate change to motivate an increase in the really big, Earth-shattering earthquakes and, as mentioned above, whether or not it can be justifiably implicated in the great seismic catastrophes of the past seven years. The only possible climate influence, however, is sea level rise. Not only is the rate of rise currently very small but, as noted in the previous chapter, rising sea levels act to stabilize the offshore subduction zone faults that play host to the world's biggest quakes. This effectively knocks on the head any idea that human activities, through climate change, could have played a role in the exceptional seismic happenings of the past several years, but it does not rule out the possibility that a warmer world may also be a more seismically active one. The aforementioned work of US geophysicists Luttrell and Sandwell shows that rising sea levels may shift the seismic focus at subduction zones landwards, promoting deeper earthquakes inland. While reducing any tsunami threat, such a migration is likely to make the shaking associated with such quakes more destructive as their epicentres would be closer to towns and cities along the coast. At the same time, laterally moving transform faults, such as California's San Andreas, and the Haitian fault whose rupture resulted in the loss of more than 250,000 lives in 2010, may generate earthquakes more easily as they are 'unclamped' due to rising sea levels bending the lithosphere where ocean meets land.

263

One of the biggest unknowns in the context of our planet's seismic future relates to the great ice sheets of Greenland and West Antarctica, where the load exerted by ice is far greater than anywhere else and where melting is occurring most rapidly. These regions are seismically quiet at present, but is this situation likely to continue as ice mass is lost at ever greater rates or will increasing numbers of earthquakes start to shake these icebound enclaves? As at Iceland's Vatnajökull Ice Cap, the lithosphere beneath Greenland is responding to shrinking of its ice cover by bouncing back upwards—and fast. According to a study by Tim Dixon of the University of Miami, and others, continued rapid ice loss is causing Greenland to rise at an accelerating rate. On the basis of data gathered from GPS stations in Greenland, Iceland and further afield, Dixon and his colleagues show that, in places, the rate of bounce-back is as high as a couple of centimetres a year and suggest that this could have doubled as soon as 2025. This may sound minor, but it is astonishing that we are seeing any uplift at all at this stage, and a clear signal of just how rapidly Greenland is divesting itself of its thick mantle of ice. Undoubtedly, the observed elevation is only a forerunner of a far greater and more widespread uplift to come, and there is certainly a long way to go. The ice is up to 2 km thick in some areas, contributing towards the highest point on the island being nearly 3700 m above sea level. The colossal weight of the ice, however, has forced down the centre of Greenland so much that it forms a basin up to 300 m *below* sea level.

Andrea Hampel of Germany's Ruhr University Bochum, and colleagues, suggest that, as well as depressing the land surface beneath, the unimaginable weight exerted by the overlying ice in Greenland and Antarctica is also enforcing the seismic quiet that currently reigns across the polar ice sheets. Not only do they expect an increase in the frequency of earthquakes in these regions as they continue to lose ice mass, but they speculate that we may see the first signs of this

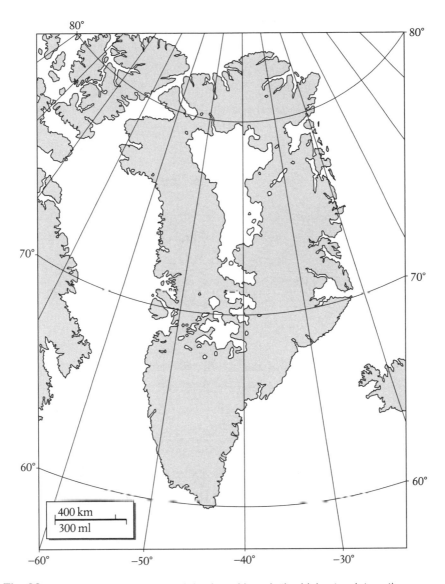

Fig 28. Such is its staggering weight that although the highest point on the surface of the Greenland Ice Sheet is nearly 3700 m above sea level, the huge mass of the ice has forced down the centre of Greenland so much that it forms a basin up to 300 m *below* sea level.

within decades. While Greenland and Antarctica can hardly be described as densely populated, future large earthquakes in these regions could end up having a surprisingly long reach if they propagate tsunamis. This is of particular concern in the case of Greenland, as the densely populated coastlines of the North Atlantic could be threatened.

Slip, sliding away

There is no reason to suspect the existence of major submarine faults around the margins of Greenland of the type that triggered the devastating tsunamis of 2004 and 2011. While the nearest plate boundary, the Mid-Atlantic Ridge, parallels the east coast not too far away, this is a constructive plate margin where new crust is being created rather than subducted, and associated earthquakes are small to middling in size. Go back 8000 years, however, and the great Storegga Slide reminds us that earthquake activity driven by the bouncing back of Scandinavia after it had divested itself of a 3 km thick ice sheet was perfectly sufficient to dislodge this vast pile of sediment. The spectacular faults exposed in Lapland provide testimony to huge post-glacial rebound quakes of magnitude 8 or more. A repeat of such seismic events, if or when the Greenland Ice Sheet is lost, clearly has the potential to promote the formation of giant submarine landslides, but only if the sediment is available. The problem is that, at the moment, we don't really know enough about the nature of the offshore zone surrounding this vast island. It is probable that this will change as the hydrocarbon companies, already climbing over one another to get at the putative reserves in the region, move in to undertake their surveys of the sea floor and the coveted treasures beneath. In the meantime, the likelihood of a future seismic rebirth for Greenland, as the ice of thousands of millennia sloughs off, has to be regarded a fair one, and the threat of resulting tsunamis triggered by submarine landslides one that needs to be considered.

In addition to promoting more in the way of earthquakes, accelerating uplift of Greenland presents another potential threat. Sea levels may be on the rise around the world, but it is quite possible—even as its melting ice contributes to this rise—that Greenland could pop up even faster. At the moment, some parts of the coast are rising six times faster than ocean levels. If this trend continues, then as the coastal margins of the island rise so the downward pressure exerted by the sea on offshore sediments will be steadily reduced. This in turn could encourage the dissociation of any gas hydrate deposits held therein. Not only would this release methane into the atmosphere, but it could also destabilize the sediment, triggering submarine slides capable of driving tsunamis out into the North Atlantic. Whether or not higher sea temperatures will promote a general increase in submarine landslides around the world due to hydrate instability, bringing with it the more widespread menace of tsunamis, is a moot point. The general consensus, at the moment, seems to be that sea level rise will act to stabilize most gas hydrates, which are stored so far beneath the sea floor that it would take centuries at least for the destabilising effect of increased ocean warmth to work its way down far enough.

If marine gas hydrate breakdown is going to be a problem anywhere as the world warms, it is likely to be around Greenland and elsewhere in the polar regions where temperatures are rising faster than anywhere else and where the cold conditions permit hydrates to be stable at much shallower depths and closer to the sea floor surface. Such deposits may already, in fact, be in a precarious position, saved from spontaneous breakdown only by a frozen carapace of submarine permafrost, whose prospects for long-term survival are poor. Monitoring by Russian scientists suggests that the permafrost shell is already starting to break up in places, releasing millions of tonnes of methane into the atmosphere of the Arctic. This could, however, be just the tip of the iceberg, with up to 1.4 trillion tonnes of gas hydrate and methane gas suspected of being trapped beneath the submarine permafrost in the region. Especially worrying is the observation that up to 10 per cent of this area is now being punctured by so-called *taliks*; areas of thawed permafrost that provide avenues for the ready escape of methane and opportunities for warmth to penetrate deep into the frozen hydrate beneath. This is a recipe for a climate catastrophe. Natalia Shakhova of the University of Alaska's International Arctic Research Centre, and her co-workers, are concerned that up to 50 billion tonnes of methane could be released abruptly and without warning from the Arctic sea bed, pushing up the methane concentration of the atmosphere 12-fold virtually overnight and driving cataclysmic warming. This, in turn, would likely lead to further methane release as permafrost on land thawed rapidly to release its store, and as wetlands—already big methane producers—spewed out even more.

The wafer-thin mint

In the course of drawing together and evaluating all the evidence, I have been persuaded that it would be remarkable if unmitigated anthropogenic climate change failed to elicit a conspicuous and salient response from the surface and interior of our planet, even if it did not fully, as it were, 'reawaken the giant'. Adding to my conviction is the fact that evidence is building for all sorts of potentially hazardous geological phenomena being extremely sensitive to even the smallest changes in their environment. I wrote earlier about volcanic eruptions being triggered or modulated by tidal stresses, by rainfall, by tiny changes in local sea level and by the minuscule variations in the shape of our planet over the course of a year. I have discussed how earthquakes can be induced by infinitesimally small variations in the rate at which our planet spins and I have introduced the idea that if a fault is critically poised, the pressure of a handshake could be sufficient to trigger its rupture and set the ground shaking. And there is much more in this vein. In Taiwan, falls in air pressure associated with typhoons crossing the island have been shown to be sufficient to promote movement on faults deep beneath the surface. Earthquake activity in parts of Japan can be linked to loading and unloading by winter snowfall, while small-scale slip and tremor on the great Cascadia Subduction Zone have been correlated with ocean tides. In Germany the timing of small earthquakes is controlled by changes in rainfall, and in Colorado, daily variations in atmospheric pressure modulate the amount of slip on the Slumgullion landslide.

Such evidence adds great weight to the idea that many potentially hazardous geological systems may be teetering on the edge of stability. As demonstrated by the bloated, full-to-bursting, Mr Creosote, in the infamous Monty Python's *Meaning of Life* sketch, who—after a gigantic, multi-course dinner—bodily exploded after accepting a final 'wafer-thin mint', a tiny nudge, in these circumstances, may initiate a reaction out of all proportion to the size of the trigger the pressure of a handshake analogy comes once again to mind. Substitute anthropogenic climate change for the wafer-thin mint and the Monty Python team may well have provided us with a metaphor for how the solid Earth will respond in the coming century and thereafter.

End game

Ultimately, the scale and extent of any response from the solid Earth to anthropogenic climate change is dependent upon the degree to which we are successful in reducing the ballooning greenhouse gas burden arising from our civilisation's increasingly polluting activities. So far, it has to be said, there is little cause for optimism in this regard. Our response to accelerating climate change continues to be consistently asymmetric, in the sense that it is far below the level that the science says is needed if we are to have any chance of avoiding all-pervasive devastating consequences.

Through our climate-changing activities we are loading the dice in favour of increased geological mayhem at a time when we can most do without it. Unless there is a dramatic and completely unexpected turnaround in the way in which the human race manages itself and the planet, then future prospects for our civilisation look increasingly grim. At a time when an additional 220,000 people are lining up at the global soup kitchen each and every night; when energy, water, and food resources are coming under ever-growing pressure; and when the debilitating effects of anthropogenic climate change are insinuating themselves increasingly into every nook and cranny of our world and our lives, the last thing we need is for the giant beneath our feet to reawaken.

SELECTED SOURCES AND FURTHER READING

1. The storm after the calm

Snowball Earth

Macdonald, F. A. et al. 2010. Calibrating the Cryogenian. *Science* **327**, 1241–3.

Mass extinctions

Hallam, A. 2005. *Catastrophes and Lesser Calamities*. Oxford University Press. Oxford. 240pp.

Schulte, P. et al. 2010. The Chicxulub asteroid impact and mass extinction at the Cretaceous—Paleogene boundary. *Science* **327**, 1214–18.

Highest atmospheric carbon dioxide levels for 15 million years

Tripati, A. K. et al. 2009. Coupling of CO_2 and ice sheet stability over major climate transitions of the last 20 million years. *Science* **326**, 1394–7.

Early human influence on the climate

Doughty, C. E. et al. 2010. Biophysical feedbacks between the Pleistocene megafauna extinction and climate: the first human-induced global warming? *Geophysical Research Letters* **37**, doi:10.1029/2010GL043985.

Ruddiman, W. F. 2005. How did humans first alter global climate. *Scientific American*, March, **292** (3), 46–53.

The rate of recent and current global warming

Hansen, J. et al. 2010. Global Surface Temperature Change. *Reviews in Geophysics* **48**, doi:10.1029/2010RG000345.

The IPCC Fourth Assessment view

IPCC 2007. *Climate Change 2007—The Physical Science Basis.* Working Group 1 Contribution to the Fourth Assessment Report. Cambridge University Press, Cambridge. 1008pp (and Working Group II and III volumes on, respectively, *Impacts, Adaptation and Vulnerability* and *Mitigation*).

The scale and speed of future warming

Betts, R. A. et al. 2011. When could global warming reach 4°C? *Philosophical Transactions of the Royal Society A.* **369**, 67–84.

Melting ice sheets and rising sea levels

Rignot, E. et al. 2008. Acceleration of the contribution of the Greenland and Antarctic ice sheets to sea-level rise. *Geophysical Research Letters* **38**, doi:10.1029/2011GL046583.

Velicogna, I. & Wahr, J. 2006. Measurements of time-variable gravity show mass loss in Antarctica. *Science* **311**, 1754–6.

Speeding-up of the hydrological cycle

Syed, T. et al. 2010. Satellite-based, global-ocean, mass balance estimates of interannual variability and emerging trends in continental freshwater discharge. *Proceedings of the National Academy of Sciences* **107**, 17916–21.

Future emissions scenarios and 'dangerous' climate change

Anderson, K. and Bows, A. 2011. Beyond 'dangerous' climate change: emission scenarios for a new world. *Philosophical Transactions of the Royal Society A.* **369**, 20–44.

The future climate change threat in its entirety

Hansen, J. 2009. *Storms of my Grandchildren*. Bloomsbury. London. 336pp.

2. Once and future climate

The Palaeocene–Eocene Thermal Maximum

Zeebe, R. E. et al. 2009. Carbon dioxide forcing alone insufficient to explain Palaeocene—Eocene Thermal Maximum warming. *Nature Geoscience* **2**, 576–80.

Gas hydrates and the PETM

Dunkley Jones, T. et al. 2010. A Palaeogene perspective on climate sensitivity and methane hydrate instability. *Philosophical Transactions of the Royal Society A.* **368**, 2395–415.

Sluijs, A. et al. 2007. Environmental precursors to rapid light carbon injection at the Palaeocene/Eocene boundary. *Nature* **450**, 1218–22.

A volcanic trigger for the PETM

Storey, M. et al. 2007. Palaeocene–Eocene Thermal Maximum and the opening of the North Atlantic. *Science* **316**, 587–9.

Carbon uptake in today's oceans

Khatiwala, S. et al. 2009. Reconstruction of the history of anthropogenic CO_2 concentrations in the ocean. *Nature* **462**, 346–50.

The climate of the Middle Miocene

Henrot, A.-J. et al. 2010. Effects of CO_2 continental distribution, topography and vegetation changes on the climate at the Middle Miocene: a model study. *Climate of the Past* **6**, 675–94.

Ice sheets and sea levels in the Pliocene

Dwyer, G. S. et al. 2009. Mid-Pliocene sea level and continental ice volume based on coupled benthic Mg/Ca palaeotemperatures and oxygen isotopes. *Philosophical Transactions of the Royal Society A.* **367**, 157–68.

The role of mountain building in cooling the Earth

Raymo, M. E. and Ruddiman, W. F. 1992. Tectonic forcing of Late climate. *Nature* **359**, 117–22.

The climate of the Eemian

Otto-Bliesner, B. L. et al. 2006. Simulating Arctic climate warmth and icefield retreat in the last interglacial. *Science* **311**, 1751–3.

Overpeck, J. T. et al. 2006. Palaeoclimatic evidence for future ice-sheet instability and rapid sea-level rise. *Science* **311**, 1747–50.

The Younger Dryas

Mullins, H. T. et al. (in press). Stable isotope evidence for Younger Dryas-Holocene climate instability, Lough Gallun, County Clare, western Ireland. Irish Journal of Earth Sciences.

The 8.2ka event

Hijma, M. P. and Cohen, K. M. 2010. The timing and magnitude of the sea-level jump preluding the 8200 ky event. *Geology* **38**, 275–8.

The climate and solar activity

Bond, G. et al. 2001. Persistent solar influence on North Atlantic climate during the Holocene. *Science* **294**, 2130–6.

3. Nice day for an eruption

The 1783 Lakagígar (Iceland) eruption

Grattan, J. et al. 2003. Illness and elevated human mortality in Europe coincident with the Laki fissure eruption. In: *Volcanic Degassing* (eds. Oppenheimer, C. et al.). Geological Society, London, Special Publications, **213**, 401–14.

Thordarson, T. and Self, S. 2003. Atmospheric and environmental effects of the 1783–84 Laki eruption: a review and reassessment. *Journal of Geophysical Research* **108**, doi:10.1029/2001JD002042.

The increase in Quaternary volcanic activity

Kennett, J. P. and Thunell, R. C. 1975. Global increase in Quaternary explosive volcanism. *Science* **187**, 497–502.

Ice Age and volcanism in Iceland

MacClennan, J. et al. 2002. The link between volcanism and deglaciation in Iceland. *Geochemistry, Geophysics, Geosystems* **3**, doi:10.1029/2001GC000282.

Quaternary volcanic eruptions in France and Germany

Nowell, D. A. G. et al. 2006. Episodic Quaternary volcanism in France and Germany. *Journal of Quaternary Science* **21**, 645–75.

Sea-level change and volcanism

McGuire, W. J. et al. 1997. Correlation between rate of sea-level change and frequency of explosive volcanism in the Mediterranean. *Nature* **473**, 473–6.

Alaska's Pavlof volcano

McNutt, S. R. 1999. Eruptions of Pavlof volcano, Alaska, and their possible modulation by ocean load and tectonic stresses: re-evaluation of the hypothesis based upon new data from 1984–1998. *Pure and Applied Geophysics* **155**, 701–12.

Earth tide triggering of volcanic eruptions

Mauk, F. J. and Johnston, M. J. S. 1973. On the triggering of volcanic eruptions by tides. *Journal of Geophysical Research* **78**, 3356–62.

The volcano season

Mason, B. G. et al. 2004. Seasonality of volcanic eruptions. *Journal of Geophysical Research* **109**, doi:10.1029/2002JB002293.

The Earth's annual deformation cycle

Blewitt, G. et al. 2001. A new global mode of Earth deformation: seasonal cycle detected. *Science* **294**, 2342–5.

The Toba super-eruption and climate

Rampino, M. R. and Self, S. 1993. Climate—volcanism feedback and the Toba eruption of 74,000 years ago. *Quaternary Research* **40**, 269–80.

Robock, A. et al. 2009. Did the Toba volcanic eruption of ~74 ka BP produce widespread glaciation? *Journal of Geophysical Research* **114**, doi:10.1029/2008JD011652.

A new idea linking ice ages, volcanoes and carbon dioxide

Huybers, P. and Langmuir, C. 2009. Feedback between deglaciation, volcanism and atmospheric CO_2. *Earth and Planetary Science Letters* **286**, 479–91.

4. Bouncing back

The 1356 Basel Earthquake

Research Management Solutions 2006. The 1356 Basel Earthquake: 650 Year Retrospective. RMS. London. 12pp.

Ice sheets and earthquakes

Stewart, I. S. et al. 2000. Glacioseismotectonics: ice sheets, crustal deformation and seismicity. *Quaternary Science Reviews* **19**, 1367–89.

The 1811–12 New Madrid earthquake

Hough, S. E. and Page, M. 2011. Towards a consistent model for strain accrual and release for the New Madrid Seismic Zone, central United States. *Journal of Geophysical Research* **116**, doi:10.1029/2010JB007783.

Deglaciation as an earthquake trigger

Muir-Wood, R. 2000. Deglaciation seismotectonics: a principal influence on intraplate seismogenesis at high latitudes. *Quaternary Science Reviews* **19**, 1399–411.

Wu, P. and Johnston, P. 2000. Can deglaciation trigger earthquakes in North America? *Geophysical Research Letters* **27**, 1323–6.

The 1967 Koyna earthquake

Talwani, P. 1995. Speculation on the causes of continuing seismicity near Koyna reservoir, India. *Pure & Applied Geophysics* **145**, 167–74.

Reservoir-induced seismicity

Talwani, P. 1997. On the nature of reservoir-induced seismicity. *Pure & Applied Geophysics* **150**, 473–92.

Draining of glacial lakes as an earthquake trigger

Karow, T. and Hampel, A. 2010. Slip-rate variations on faults in the Basin-and-Range Province caused by regression of Late Pleistocene Lake Bonneville and Lake Lahontan. *International Journal of Earth Sciences* **99**, 1941–53.

Ringrose, P. S. 1989. Palaeoseismic(?) liquefaction event in late Quaternary lake sediment at Glen Roy, Scotland. *Terra Nova* **1**, 57–62.

Ice-mass loss and earthquake activity in Alaska

Doser, D. I. et al. 2007. Seismicity of the Bering Glacier region and its relation to tectonic and glacial processes. *Tectonophysics* **439**, 119–27.

Sauber, J. M. and Molnia, B. F. 2004. Glacier ice mass fluctuations and fault instability in tectonically active Southern Alaska. *Global Planetary Change* **42**, 279–93.

The Zipingpu reservoir and the 2008 Wenchuan earthquake

Klose, C. 2008. The 2008 M7.9 Wenchuan earthquake—result of local and abnormal mass imbalances? *EOS Transactions AGU* **89** (53), Fall Meeting Supplement. Abstract U21C-08.

5. Earth in motion

The formation of the Valle del Bove

Deeming, K. R. et al. 2010. Climate forcing of volcano lateral collapse: evidence from Mount Etna, Sicily. *Philosophical Transactions of the Royal Society, A.* **368**, 2559–77.

Pareschi, M. T. et al. 2006. Lost tsunami. *Geophysical Research Letters* **33**, doi:10.1029/2006GL027790.

Rapid climate change and collapsing volcanoes

Capra, L. 2006. Abrupt climatic changes as triggering mechanisms of massive volcanic collapses. *Journal of Volcanology & Geothermal Research* **155**, 329–33.

Collapsing volcanoes and tsunamis

McGuire, W. J. 2006. Lateral collapse and tsunamigenic potential of marine volcanoes. In: Troise, C et al. (eds). *Mechanisms of Activity and Unrest at Large Calderas*. Geological Society, London, Special Publications, **269**, 121–40.

Instability at ocean island volcanoes and climate change

Keating, B. H. and McGuire, W. J. 2004. Instability and structural failure at volcanic ocean islands and continental margins and the climate change dimension. *Advances in Geophysics* **47**, 175–271.

Giant Hawaiian landslides

Moore, J. G. et al. 1989. Prodigious submarine landslides on the Hawaiian Ridge. *Journal of Geophysical Research* **94**, 17465–84.

Megatsunami deposits in Hawaii

McMurtry, G. M. 2004. Megatsunami deposits on Kohala volcano, Hawaii, from flank collapse at Mauna Loa. *Geology* **32**, 741–4.

A future megatsunami sourced in the Canary Islands

Ward, S. N. and Day, S. J. 2001. Cumbre Vieja volcano—potential collapse and tsunami at La Palma, Canary Islands. *Geophysical Research Letters* **28**, 397–400.

Slow cracking and the Vajont dam disaster

Kilburn, C. R. J. and Petley, D. N. 2003. Forecasting giant, catastrophic slope collapse: lessons from Vajont, northern Italy. *Geomorphology* **54**, 21–32.

The Storegga landslide and tsunami

Bondevik, S. J. I. et al. 2005. The Storegga Slide tsunami—comparing field observations with numerical simulations. *Marine & Petroleum Geology* **22**, 195–208.

Bryn, P. et al. 2005. Explaining the Storegga Slide. *Marine & Petroleum Geology* **22**, 11–19.

Submarine landslides, tsunamis and climate change

Tappin, D. R. 2010. Submarine mass failures as tsunami sources: their climate control. *Philosophical Transactions of the Royal Society, A.* **368**, 2417–34.

Gas hydrates and submarine landslides

Maslin, M. et al. 2010. Gas hydrates: past and future geohazard. *Philosophical Transactions of the Royal Society, A.* **368**, 2369–93.

6. Water, water, everywhere

Origin of the Earth's water

Genda, H. and Ikoma, M. 2008. Origin of the ocean on the Earth: early evolution of water D/H in a hydrogen-rich atmosphere. *Icarus* **194**, 42–52.

Oceans in Earth's earliest history

Watson, E. B. and Harrison, T. M. 2005. Zircon thermometer reveals minimum melting conditions on earliest Earth. *Science* **308**, 841–4.

Sea-level change in geological time

Haq, B. U. et al. 2008. A chronology of Palaeozoic sea-level change. *Science* **322**, 64–8.

Catastrophic flooding of the Mediterranean

Garcia-Castellanos, D. et al. 2009. Catastrophic flood of the Mediterranean after the Messinian salinity crisis. *Nature* **462**, 778–82.

Flooding of the Black Sea Basin

Giosan, L. et al. 2009. Was the Black Sea catastrophically flooded in the early Holocene? *Quaternary Science Reviews* **28**, 1–6.

Linking fault movement and the length of day

Wang, Qing-Liang et al. 2000. Decadal correlation between crustal deformation and variation in length of the day of the Earth. *Earth Planets Space* **52**, 989–92.

The Late Pleistocene Mediterranean mega-turbidite

Rothwell, R. G. et al. 1998. Low-sea-level emplacement of a very large Late Pleistocene 'megaturbidite' in the western Mediterranean Sea. *Nature* **392**, 377–80.

Giant submarine landslides off the Amazon delta

Maslin, M. et al. 1998. Sea-level- and gas-hydrate-controlled catastrophic sediment failures of the Amazon Fan. *Geology* **26**, 1107–10.

A volcanic and tectonic response to large changes in sea level

Nakada, M. and Yokose, H. 1992. Ice age as a trigger of active Quaternary volcanism and tectonism. *Tectonophysics* **212**, 321–9.

Low sea level and eruptions of Pantelleria

Wallmann, P. C. et al. 1988. Mechanical models for correlation of ring fracture eruptions at Pantelleria, Strait of Sicily, with glacial sea-level drawdown. *Bulletin of Volcanology* **50**, 327–39.

Ocean loading of plate boundary faults

Luttrell, K. and Sandwell, D. 2010. Ocean loading effects on stress at near shore plate boundary fault systems. *Journal of Geophysical Research* **115**, doi:10.1029/2009JB006541.

The Cascadia subduction zone

Goldfinger, C. et al. (in press). Turbidite event history: methods and implications for Holocene paleoseismicity of the Cascadia Subduction Zone. *USGS Professional Paper 1661-F.* United States Geological Survey, Reston, Virginia. 178pp.

Catastrophic sea-level rise events

Blanchon, P. and Shaw, J. 1995. Reef-drowning during the last deglaciation: evidence for catastrophic sea-level rise and ice-sheet collapse. *Geology* **23**, 4–8.

Chapter 7. Reawakening the giant

A response from the geosphere to future climate changes

McGuire, W. J. 2010. Potential for a hazardous geospheric response to projected future climate changes. *Philosophical Transactions of the Royal Society A.* **368**, 2317–46.

Sea level during the last interglacial

Kopp, R. E. et al. 2009. Probabilistic assessment of sea level during the last interglacial stage. *Nature* **462**, 863–7.

Giant landslides in Alaska, the Caucasus and elsewhere

Huggel, C. et al. 2008. Recent extreme avalanches: triggered by climate change? *Eos* **89**, 469–84.

Mountain instability and heat waves

Huggel, C. et al. 2010. Recent and future warm extreme events and high-mountain slope stability. *Philosophical Transactions of the Royal Society A.* **368**, 2435–60.

The Greenland Ice Sheet's tipping point

Mernild, S. H. 2010. Greenland Ice Sheet surface mass-balance modelling in a 131-yr perspective, 1950–2080. *Journal of Hydrometeorology* **11**, 3–25.

Rejuvenating Vatnajökull's volcanoes

Sigmundsson, F. et al. 2010. Climate effects on volcanism: influence on magmatic systems of loading and unloading from ice mass variations, with examples from Iceland. *Philosophical Transactions of the Royal Society A*. **368**, 2519–34.

Climate change and the volcanic threat

Tuffen, H. 2010. How will melting of ice affect volcanic hazards in the twenty-first century? *Philosophical Transactions of the Royal Society A*. **368**, 2535–58.

Acceleration of melting at the poles and future sea level rise

Rignot, E. et al. 2011. Acceleration of the contribution of the Greenland and Antarctic ice sheets to sea level rise. *Geophysical Research Letters* **38**, doi:10.1029/2011GL046583.

Greenland bounces back

Jiang, Y. et al. 2010. Accelerating uplift in the North Atlantic region as an indicator of ice loss. *Nature Geoscience* **3**, 404–7.

A seismic response to climate change in Greenland and Antarctica

Hampel, A. et al. 2010. Response of faults to climate-driven changes in ice and water volumes on Earth's surface. *Philosophical Transactions of the Royal Society A*. **368**, 2501–18.

The Arctic methane 'time-bomb'

Shakhova, N. et al. 2008. Anomalies of methane in the atmosphere over the East Siberian Shelf: is there any sign of methane leakage from shallow shelf hydrates? *Geophysical Research Abstracts* **10**, EGU2008-A-01526. European Geosciences Union General Assembly 2008.

INDEX